FRACTALS

PHYSICS OF SOLIDS AND LIQUIDS

Editorial
Board: Jozef T. Devreese • *University of Antwerp, Belgium*
Roger P. Evrard • *University of Liège, Belgium*
Stig Lundqvist • *Chalmers University of Technology, Sweden*
Gerald D. Mahan • *Indiana University, Bloomington, Indiana*
Norman H. March • *University of Oxford, England*

AMORPHOUS SOLIDS AND THE LIQUID STATE
Edited by Norman H. March, Robert A. Street, and Mario P. Tosi

CHEMICAL BONDS OUTSIDE METAL SURFACES
Norman H. March

CRYSTALLINE SEMICONDUCTING MATERIALS AND DEVICES
Edited by Paul N. Butcher, Norman H. March, and Mario P. Tosi

ELECTRON SPECTROSCOPY OF CRYSTALS
V. V. Nemoshkalenko and V. G. Aleshin

FRACTALS
Jens Feder

HIGHLY CONDUCTING ONE-DIMENSIONAL SOLIDS
Edited by Jozef T. Devreese, Roger P. Evrard, and Victor E. van Doren

MANY-PARTICLE PHYSICS
Gerald D. Mahan

ORDER AND CHAOS IN NONLINEAR PHYSICAL SYSTEMS
Edited by Stig Lundqvist, Norman H. March, and Mario P. Tosi

THE PHYSICS OF ACTINIDE COMPOUNDS
Paul Erdös and John M. Robinson

POLYMERS, LIQUID CRYSTALS, AND LOW-DIMENSIONAL SOLIDS
Edited by Norman H. March and Mario P. Tosi

SUPERIONIC CONDUCTORS
Edited by Gerald D. Mahan and Walter L. Roth

THEORY OF THE INHOMOGENEOUS ELECTRON GAS
Edited by Stig Lundqvist and Norman H. March

A Continuation Order Plan is available for this series. A continuation order will bring delivery of each new volume immediately upon publication. Volumes are billed only upon actual shipment. For further information please contact the publisher.

FRACTALS

Jens Feder

Department of Physics
University of Oslo
Oslo, Norway

PLENUM PRESS • NEW YORK AND LONDON

Library of Congress Cataloging in Publication Data

Feder, Jens.
  Fractals.

  (Physics of solids and liquids)
  Bibliography: p.
  Includes index.
  1. Fractals. I. Title. II. Series.
QA447.J46   1988   516′.13   87-31447
ISBN 0-306-42851-2

© 1988 Plenum Press, New York
A Division of Plenum Publishing Corporation
233 Spring Street, New York, N.Y. 10013

Printed in the United States of America

for Liv,
   Heidi, and Brummen

Foreword

This lovely little book will take off and fly on its own power, but the author has asked me to write a few words, and one should not say no to a friend. Specific topics in fractal geometry and its applications have already benefited from several excellent surveys of moderate length, and gossip and preliminary drafts tell us that we shall soon see several monographic treatments of broader topics. For the teacher, however, these surveys and monographs are not enough, and an urgent need for more helpful books has been widely recognized. To write such a book is no easy task, but Jens Feder meets the challenge head on. His approach combines the old Viking's willingness to attack many difficulties at the same time, and the modern Norwegian's ability to achieve fine balance between diverging needs. I owe him special gratitude for presenting the main facts about R/S analysis of long-run dependence; now a wide scientific public will have access to a large group of papers of mine that had until this day remained fairly confidential.

Last but not least, we are all grateful to Jens for not having allowed undue personal modesty to deprive us of accounts of his own group's varied and excellent work. He did not attempt to say everything, but what he said is just fine.

Benoit B. Mandelbrot
Physics Department, IBM Thomas J. Watson Research Center
Yorktown Heights, New York 10598
and Mathematics Department, Yale University
New Haven, Connecticut 06520

Preface

This book grew out of research on phase-transitions, on the aggregation of immunoglobulins and recently on the viscous fingering of fluid displacement in porous media. These research subjects represent examples of the general question: How is the microscopic behavior related to what we observe on the macroscopic scale? I now feel that fractals, relating geometry on different scales, are essential for the description and understanding of this relation. My friend Torstein Jøssang and I have for many years attempted to gain insight into the connection between microscopic physics and macroscopic phenomena by means of experiments, theory and computer simulations. Much of what I know I have learned through association with Torstein. He also contributed to early lecture notes and reports on fractals. However, he felt that I should write a monograph. We have also had the benefit of working with many talented students.

This book contains some of the topics I found particularly interesting and useful in teaching and in our research. Many more ideas and interesting facets of fractals are found in Mandelbrot's books and in the rapidly growing research literature. I would like to apologize to all colleagues whose works have not been cited as my aim has been to write an introduction that may be useful for those who want to use fractals and not to write an exhaustive review.

The interesting phenomena that occur in the displacement of fluids in porous media became a focus of research in our cooperation with Den norske stats oljeselskap a/s (Statoil). This cooperation has exposed us to many questions of practical interest that have quickly evolved into questions of basic research. Our research effort and our students have benefited from the generous support of VISTA, a cooperative research effort between

the Norwegian Academy of Science and Letters and Statoil, initiated by
director Henrik Ager-Hanssen.

Dr. Per Stokke at Statoil raised many interesting questions and asked
for several reports on the possible applications of fractals relating to geological, geochemical and other subjects of direct interest in oil exploration.
These reports started the process of actually writing this book. Teaching
a course to our students on the application of fractals pushed the project
along and raised many new questions.

Amnon Aharony has encouraged me in writing this book and I have
learned much from his many constructive remarks. Ivar Giaever read a
preliminary version and made many penetrating comments and suggestions
that I have tried to incorporate. I learned much from discussions with Paul
Meakin. Jens Lothe has commented on several parts of the book, and I
am also indebted to him as my thesis advisor. Harry Thomas made helpful
comments on my first writings on the subject. During the summer of 1986,
I visited Erling Pytte at the IBM Thomas J. Watson Research Center, and
he suggested many improvements to the book. Benoit Mandelbrot read
preliminary versions of my book during the summer and we had many
interesting discussions. He pointed out, with humor and patience, errors
and misunderstandings, made valuable suggestions and encouraged me in
many ways. I am grateful to him for his inspiration and help.

Jan Frøyland has contributed much to the analysis of wave-height
statistics. He has also generated many of the random translation surfaces
shown in chapter 13. Many of the students Jøssang and I have in our group
have contributed directly to this book. Knut Jørgen Måløy has with ingenu-
ity done the experiments on viscous fingering in porous media. Unni Oxaal
has made experiments on fluid displacement in micromodels of controlled
geometry. Einar Hinrichsen carried out DLA simulations and also made
many useful comments on the manuscript. Finn Boger has contributed to
the analysis of experimental results. He has also developed programs that
have generated the fractal landscapes and clouds shown in this book. Liv
Furuberg simulated percolation processes and has contributed many of the
figures in chapter 7.

Liv Feder created most of the illustrations that appear in the book.
She has helped me in many ways, and this book would not have been
written without her patience and support.

Without doubt, this book can be improved. If you have comments or suggestions, I would be pleased to have them.

Jens Feder

Oslo, Norway

Contents

Color Plates

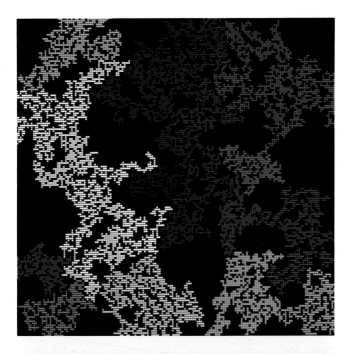

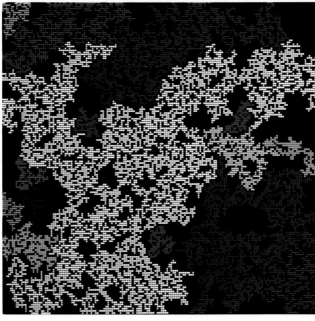

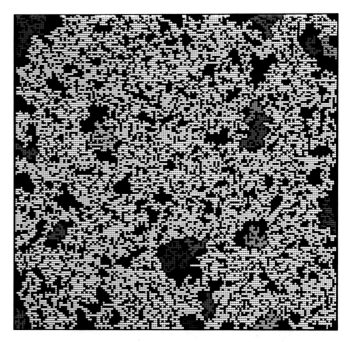

FIGURE C.1: The effect of increasing the occupation probability p on a 160×160 quadratic lattice. From top to bottom we have $p = 0.58$, 0.6 (see facing page) and 0.62 above. In each of the three figures the largest cluster is *white*. The other clusters are colored according to decreasing size with the colors *cyan, red, orange, yellow, light green, green, turquoise* and *blue* in the shades light to dark. The smallest clusters are not visible with this coloring scheme.

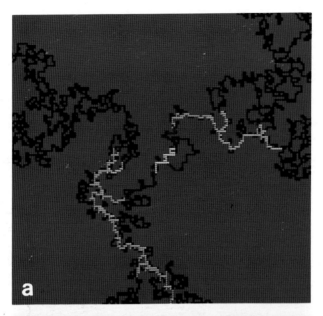

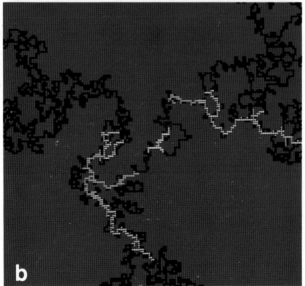

FIGURE C.2: (a) Air displacing glycerol at a high capillary number on the percolation cluster shown in figure 7.13. (b) The results of numerical simulation of fluid displacement on the same percolation cluster. The different colors represent pores invaded by air observed at successive time steps. The number of pores invaded by air is 30 – *white*, 86 – *red*, 213 – *green* and finally at breakthrough 447 – *yellow*, for both the experiment and the simulation (Oxaal et al., 1987).

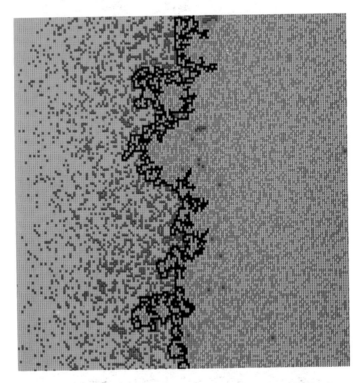

FIGURE C.3: A system of diffusing particles at large scale. The screen is 300 × 300. The particles connected to the source are *green*, empty sites connected to the sink are *turquoise*. The particles in isolated clusters (islands) are *light green*, isolated empty sites (lakes) are *dark blue*. Sites in the hull are *black*.

FIGURE C.4: Fractal landscape with $H = 0.75$ generated using a scale factor $r = \frac{1}{2}$ on a 2049 × 2049 lattice. (a) The central 1024 × 800 portion seen from above. (b) The landscape with perspective and curvature. (c)

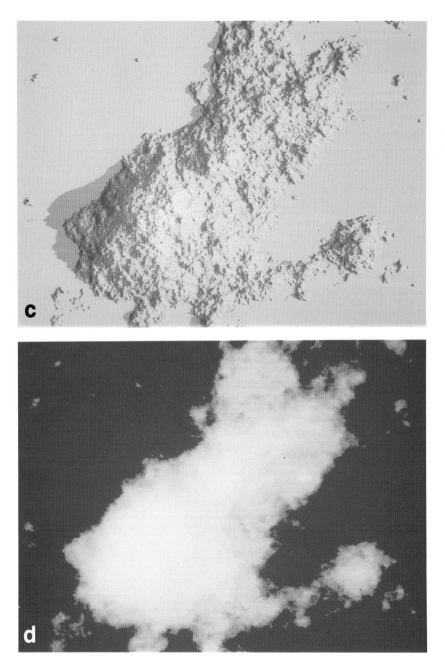

The complete landscape seen from above. (d) The complete landscape presented as 'clouds.' For a discussion see section 13.4 (Boger et al., 1987).

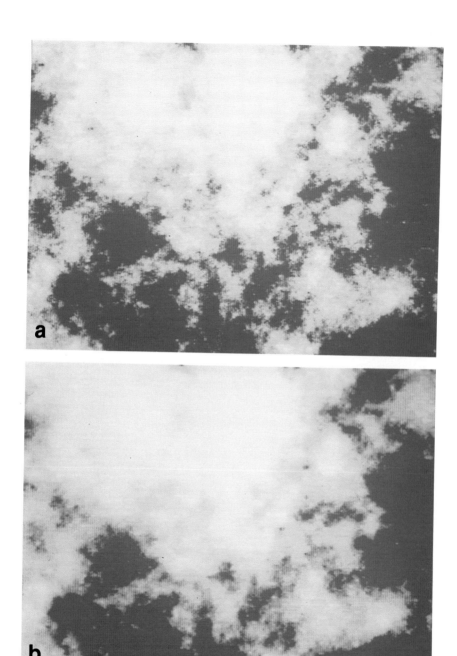

FIGURE C.5: Fractal landscapes presented as 'clouds.' (a) $H = 0.5$ landscape (see figure 13.11). (b) $H = 0.7$ landscape (see figures 13.11 and 13.13) (Boger et al., 1987).

FRACTALS

Chapter 1

Introduction

The geometry of natural objects ranging in size from the atomic scale to the size of the universe is central to the models we make in order to 'understand' nature. The geometry of particle trajectories; of hydrodynamic flow lines, waves, ships and shores; of landscapes, mountains, islands, rivers, glaciers and sediments; of grains in rock, metals and composite materials; of plants, insects and cells, as well as the geometrical structure of crystals, chemicals and proteins — in short the geometry of nature is so central to the various fields of natural science that we tend to take the geometrical aspects for granted. Each field tends to develop adapted concepts (e.g., morphology, four-dimensional spaces, texture, conformation, and dislocations) used intuitively by the scientists in that field. Traditionally the Euclidean lines, circles, spheres and tetrahedra have served as the basis of the intuitive understanding of the geometry of nature.

Mathematicians have developed geometrical concepts that transcend traditional geometry, but unfortunately these concepts have failed in the past to gain acceptance in the natural sciences because of the rather abstract and 'pedantic' presentations, and because of warnings that such geometries were 'dangerous to use'!

Benoit B. Mandelbrot, with his creative and monumental work, has generated a widespread interest in *fractal geometry* — a concept introduced by Mandelbrot himself. In particular he has presented what he has called fractals in an unusually inspiring way. His book *The Fractal Geometry of Nature* (1982) is the standard reference and contains both the elementary concepts and an unusually broad range of new and rather advanced ideas, such as multifractals, currently under active study. The pictures of synthetic landscapes look so real that they are accepted by most to be natural. The advent in recent years of inexpensive computer power and graphics has led to the study of nontraditional geometrical objects in many fields of science.

1

Mandelbrot has written a large number of scientific papers that deal with the geometry of the phenomena observed in many fields of science. He has studied the fractal geometry of price changes and salary distributions; of the statistics of errors in telephone messages; of word frequencies in written texts; of various mathematical objects and of many other subjects. He has written three books[1] on the subject that make his technical papers more accessible and that have inspired many to use fractal geometry in their own fields:

1. *Les Objets Fractals: Forme, Hasard et Dimension* (1975a).

2. *Fractals: Form, Chance, and Dimension* (1977).

3. *The Fractal Geometry of Nature* (1982).

The last book is a new edition with spectacular pictures generated using the greatly enhanced computer graphics now available.

The concept of fractals has caught the imagination of scientists in many fields and papers discussing fractals in various contexts now appear almost daily. Mandelbrot's books are remarkable in several ways. First of all, they are cross-disciplinary — he discusses trees, rivers, lungs, water levels, turbulence, economics, word frequencies and many, many other topics. He ties all these fields together with his geometrical concepts. He purposely has no introduction and no conclusion in order to stress his belief that as more work is done in this field, his ideas will reveal further insight into the geometry of nature. In fact, he only reluctantly gives a definition of the term fractal and he hastens to state that his definition is only tentative! Later he withdraws this definition. His books try to convince the reader that fractal geometry is important for the description of nature, but become elusive when the reader tries to understand the arguments in detail. Mathematical arguments are mixed with anecdotes and historical notes. Various topics are mixed throughout the book in a way that is difficult to disentangle. However, with patience the interested reader finds an unusual spectrum of good ideas, profound remarks and inspiration — these books are truly remarkable.

The most striking illustrations are in color. They show a fractal 'planet' rising over the horizon of its moon, and mountains, valleys and islands that never were. These illustrations, made by R. F. Voss, are based

---

[1]Mandelbrot (1988) is preparing a new book that also incorporates his papers of 1967, 1972, 1974, 1985 and 1986 found in the list of references, as well as his other hard-to-find papers on multifractals. It is to be Volume I of Mandelbrot's *Selecta*. Additional volumes, incorporating the other references, are anticipated and may include the papers on R/S analysis.

on algorithms that ensure the fractal nature of the landscapes. The landscapes look natural — one must believe that the fractals somehow capture the essence of the surface topography of the earth. How these fascinating pictures are made is not explained, but Mandelbrot states that

> *'implementing the shadows involves great ingenuity; one would need tomes to explain every detail. In addition the algorithm is very much influenced by the available tools, hence to duplicate this work one should have to use exactly the same computer equipment'* (page C8).

Voss has recently described in some detail the ideas he uses in the generation of his spectacular pictures (Voss 1985a,b). Some of the computer graphics involved in the creation of beautiful fractal objects is described in the proceedings of the computer graphics section of the Association for Computing Machinery (Siggraph 86 conference proceedings, *Computer Graphics*, edited by D. C. Evans and R. J. Athay, 1986). In these proceedings fascinating fractal trees created by Oppenheimer (1986) are well worth studying.

Peitgen and Richter's new book *The Beauty of Fractals* (1986) is a truly beautiful book exploring the fractal nature of iterative maps and of differential equations. Peitgen is also editing *The Art of Fractals, A Computer Graphical Introduction* to appear in 1988.

In order to gain insight into these types of landscapes we have generated some landscapes of our own using methods described in chapter 13. Anyone who tries such calculations will conclude that computational shortcuts are necessary and understand why Mandelbrot discusses many different possible schemes — *'I have also examined and compared a dozen shortcuts that are stationary, and some day I hope to publish the comparison'* (Mandelbrot, 1982). We find that simple models of landscapes and coastlines can be produced quickly with rather limited computer resources. We have learned much from this effort and recommend that the reader try to generate some as well.

In order to provide a basis for the various applications of fractals to experimental results we start out in chapter 2 with a discussion of simple fractals and of the fractal dimension. The related concept of the scaling or similarity dimension is also discussed.

In chapter 3 the fractal properties of clusters are discussed and experimental results reviewed. Aggregation of particles has been shown to produce fractal clusters. Many experiments and numerical simulations have recently explored the properties of aggregation kinetics, gelation and sedimentation. Various experimental techniques have been applied in the study of these phenomena and the use of fractal geometry has helped to rationalize large sets of experimental results. For a recent review see the paper by Meakin (1987c).

Displacement of a fluid in porous media usually results in a displacement front. If the fluid is driven by a fluid of lower viscosity the displacement front is notoriously unstable. Such fronts have been extensively studied experimentally and theoretically. It is now clear that displacement using a low-viscosity fluid or gas is mathematically analogous to the kinetics of an aggregation process which is known to produce fractal geometries. We discuss the theoretical background and experimental results in chapter 4.

Once one leaves the secure ground of conventional geometry a whole zoo of fractal dimensions appears. As a preparation we discuss the simple example of Cantor sets in chapter 5. In chapter 6 we discuss what happens when we consider physical phenomena or distributions on fractals, and we present the ideas of fractal measures and multifractals in a discussion of simple examples. We then use these ideas in a discussion of some recent experimental results on thermal convection and on the dynamics of viscous fingering.

Randomness is an essential ingredient of most natural phenomena. In chapter 7 we discuss percolation processes, which provide particularly well understood examples of random fractals. We concentrate on the fractal geometry of percolation processes and proceed to discuss experimental results obtained in fluid–fluid displacement.

Many records of observations exhibit fractal statistics which may be analyzed using the empirical R/S analysis introduced by Mandelbrot and Wallis (1968) on the basis of puzzling observations by Hurst and presented in chapter 8. This analysis gives evidence that many natural records in time are fractal. The more detailed discussion of random walks in chapter 9 provides the tools needed to understand fractal time series, and introduces the concept of fractional Brownian motion.

The analysis of the fractal structure of records in time emphasizes the need to distinguish between self-affine and self-similar fractals. It has recently become clear that great care has to be taken in order to obtain meaningful fractal dimensions for self-affine fractals. Chapter 10 discusses some of these difficulties. The discussion of the 'strategy of bold play' gives an interesting example of a self-affine curve that is directly related to the fractal measures discussed in chapter 6. As an application of the R/S statistics to a self-affine record in time I discuss in chapter 11 the wave-height statistics of ocean waves.

The perimeter–area relation for fractal objects is discussed in chapter 12. This discussion forms the basis for an understanding of Lovejoy's (1982) observation that the surface of clouds has a fractal dimension $D = 2.34$. The fractal dimension has been calculated from 'first principles' by Hentschel and Procaccia (1984), who find D in the range 2.37 to 2.41 for clouds based on a model of turbulent diffusion. Some observations on the relation between river length and drainage areas indicate that river systems are fractals, as discussed in the last section.

Fractal surfaces are the subject of chapter 13. The relation between fractal curves and surfaces is the basis for our discussion. Defining and using random fractal translation surfaces we have made drawings of fractal coastal landscapes and coastlines. The successive random addition method used by Voss provides an efficient algorithm for the generation of fractal surfaces with specified properties. We show a few examples of such surfaces.

The recent discussion of fractal dimensions based on topographic measurements of surfaces is covered in chapter 14. Also the related 'observation' of the fractal dimensions of various environmental data is presented. The surfaces of powders and other porous media have recently been shown to be fractal, and many of these results are also discussed. The fractal nature of surfaces must have implications for catalysis and for the properties of porous media. In short the fractal dimension D of a porous substance is a property of the material and must in turn determine many of its properties.

Many highly interesting and important topics are not discussed in this book. Dynamical systems, for example, provide cases in which fractals arise naturally, and the reader would be well advised to study these topics in one of the many books that are now available. Peitgen and Richter's (1986) book is a good starting point. Another theme of very active research is the dynamics of various processes that occur in fractal geometries. For example, the electrical conductivity, noise and mechanical properties of percolating systems provide important examples of multifractal behavior. A theme that deserves further discussion is the random walk in higher dimensions on fractals or with fractal steps. The list goes on and on. Why not discuss the fractal structure of turbulence, the formation of galaxies and the distribution of fractures? Why not indeed — the research activity in this field is huge and growing. Many interesting topics have been discussed at a number of conferences: Family and Landau (1984); Pynn and Skjeltorp (1985); Stanley and Ostrowsky (1985); Pietronero and Tosatti (1986); Pynn and Riste (1987). The selection of the themes presented here has been guided mostly by my own research interests and by the questions and demands of my students. The subject is changing very fast, with interesting contributions and applications being published daily. Also, many of the themes discussed in Mandelbrot's books have not even been touched upon here. His books remain the main reference and the reader of this book should explore them for new and interesting ideas.

Chapter 2

The Fractal Dimension

2.1   The Coast of Norway

How long is the coast of Norway? Take a look at figure 2.1. On the scale of the map the deep fjords on the western coast show up clearly. The details encountered moving northeast along the coast from the southern tip are more difficult to resolve, but I can assure you that the maps I use when sailing in that area show structures quite similar to those of the west coast. In fact when sailing there you find rocks, islands, bays, faults and gorges that look much the same but do not show up even on my detailed maps. Before answering the question we have to decide on whether the coast of the islands should be included. And what about the rivers? Where does the fjord stop being a fjord and become a river? Sometimes this is an easy question and sometimes not. However, even after we have decided on all these questions we have a difficulty. I could walk a divider with an opening corresponding to δ km along the coastline on the map and count the number of steps, $N(\delta)$, needed to move from one end of the map to the other. Being in a hurry I would choose such a large opening of the divider that I would not have to bother about even the deepest fjords and estimate the length to be $L = N(\delta) \cdot \delta$. If objections were raised I would use a somewhat smaller opening δ and try again. This time the large fjords would contribute to the measured length but the southeastern coast would still be taken in a few steps. For a really serious discussion I would have to get the kind of maps neighbors use when they settle questions of where the fence should go, or how far up the river the fishing rights extend. Clearly there is no end to this line of investigation. Every time we increase the resolution we find an increase in the measured length of the coastline. Also in using the divider we have problems with the islands and rivers. An alternative method of measuring the length of the coastline is to cover the

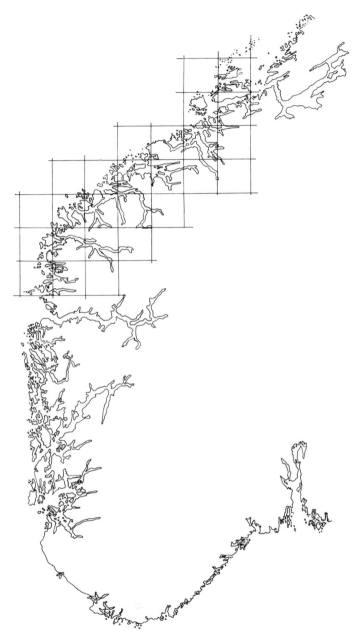

FIGURE 2.1: The coast of the southern part of Norway. The outline was traced from an atlas and digitized at about 1800×1200 pixels. The square grid indicated has a spacing of $\delta \sim 50$ km.

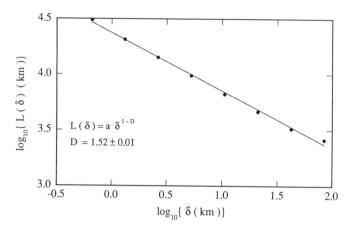

FIGURE 2.2: The measured length of the coastline shown in figure 2.1, as a function of the size δ of the $\delta \times \delta$ squares used to cover the coastline on the map. The straight line in this log–log plot corresponds to the relation $L(\delta) = a \cdot \delta^{1-D}$, with $D \simeq 1.52$.

map with a grid, as indicated in a corner of figure 2.1. Let the square cells of the grid be $\delta \times \delta$. The number $N(\delta)$ of such squares needed to cover the coastline on the map is roughly equal to the number of steps used when walking a divider with an opening δ along the coast. Decreasing δ again gives a large increase in the number of cells needed to cover the coastline. If the coast of Norway has a well defined length L_N, then we expect that the number of steps taken using the divider, or the number of square cells needed to cover the coast, $N(\delta)$, will be inversely proportional to the δ, so that $L(\delta) = N(\delta) \times \delta$ approaches a constant L_N as we make δ smaller and smaller. However, this is not the case.

Figure 2.2 shows how the measured length increases as the 'yardstick' length, δ, is reduced. This log–log plot shows that the measured length of the coastline shows no sign of reaching a fixed value as δ is reduced. In fact, the measured length is nicely approximated by the formula

$$L(\delta) = a \cdot \delta^{1-D} . \tag{2.1}$$

For an ordinary curve we would expect a to be L_N, at least for small enough δ, and the exponent D should be equal to one. We find, however, that $D \simeq 1.52$. The coastline is a fractal with a fractal dimension D. We discuss this in more detail in section 2.3.

In Mandelbrot's (1982) book is a chapter entitled 'How Long Is the

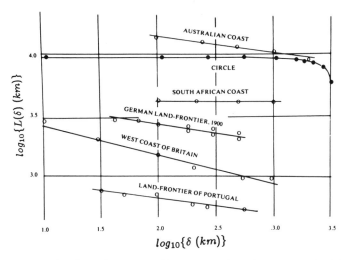

FIGURE 2.3: The length of coastlines as a function of yardstick length (Mandelbrot, 1982).

Coast of Britain?' He analyzed data collected by Richardson, and figure 2.3 is a reproduction of his figure showing the apparent length of various coastlines and boundaries (Mandelbrot, 1967). They all fall on straight lines in the log–log plot. The slope of the lines in this log–log plot is $1 - D$, where D is the fractal dimension of the coastline. The coast of Britain has $D \sim 1.3$. Mandelbrot added the results for a circle in his figure, and he finds $D_{\text{circle}} = 1$, as was expected.

2.2   The Schwarz Area Paradox

Measuring an area is not always easy in practice. Consider the surface of the cylinder (radius R and height H) illustrated in figure 2.4; its area is $A = 2\pi RH$. However, if we try to measure the surface area of a given cylinder in practice using rulers, we would have to *triangulate* the surface in some way, for instance as shown in figure 2.4. We divide the surface into m bands and n sectors as indicated in the figure, and obtain an estimate of the surface area as the sum A_Δ of the areas of all the small triangles. By making this division finer and finer, that is, letting $n \to \infty$ and $m \to \infty$, we expect that $A_\Delta \to A$. This is not always correct. The area of all the

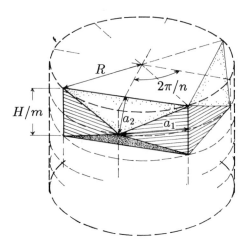

FIGURE 2.4: The vertical surface of a cylinder of radius R and height H is $2\pi RH$. The surface is approximated by a triangulation as illustrated.

triangles may be written as

$$
\begin{aligned}
A_\Delta &= \pi RH \tfrac{2n}{\pi} \sin \tfrac{\pi}{2n} \left\{ 1 + \cos \tfrac{\pi}{2n} \sqrt{1 + \left(\tfrac{R}{H}\right)^2 \tfrac{\pi^4 m^2}{n^4} \left(\tfrac{2n}{\pi} \sin \tfrac{\pi}{2n}\right)^4} \right\} \\
&\xrightarrow[n \to \infty]{} \pi RH \left\{ 1 + \sqrt{1 + (R/H)^2 (\pi^2 m/n^2)^2} \right\} .
\end{aligned}
$$

$$(2.2)$$

Here the first term corresponds to triangles of the type marked a_1 in the figure. The second term with the square-root corresponds to triangles of the type marked a_2. We see that if $m/n^2 \to 0$, as both m and n increase, then we find indeed that the triangulated area approaches the expected result. However, if we choose to use a triangulation where $m = \lambda n^2$, then we find that $A_\Delta > A$, and we may in fact get arbitrarily large values for A_Δ. If we choose $m = n^\beta$, then we find $A_\Delta \sim n^{\beta - 2}$ for $\beta > 2$. The triangulation area then diverges as we make the individual triangles smaller and smaller. Instead of getting a better approximation by a reduction of the triangle size we obtain a worse approximation. Many other ways of triangulation lead to similar problems. This is the Schwarz area paradox. For a discussion see Mandelbrot (1986). It is easy to see what happens. We find that the approximate triangulation surface becomes more and more corrugated as the ratio m/n^2 increases, and in the limit the triangles of the type marked a_2 are practically perpendicular to the cylinder surface.

One may object that we run into trouble only with a bad choice of the triangulation. However, how should one select a 'good' triangulation when we want to estimate the area of a more complex or rough surface? One finds that one is better off using the methods discussed in the next section. They work in the classical and easy case and in the more difficult cases of 'monster' curves, surfaces and volumes.

2.3   The Fractal Dimension

Mandelbrot (1982) offers the following *tentative* definition of a fractal:

> *A fractal is by definition a set for which the Hausdorff-Besicovitch dimension strictly exceeds the topological dimension* (page 15).

This definition requires a definition of the terms *set, Hausdorff-Besicovitch dimension* (D) and *topological dimension* (D_T), which is always an integer. For the present purpose we find that a rather loose definition of these terms and illustrations — using simple examples — is more useful than the more formal mathematical discussions available. In fact Mandelbrot (1986) has retracted this tentative definition and proposes instead the following[1]:

> *A fractal is a shape made of parts similar to the whole in some way.*

A neat and complete characterization of fractals is still lacking (Mandelbrot, 1987). The point is that the first definition, although correct and precise, is too restrictive. It excludes many fractals that are useful in physics. The second definition contains the essential feature that is emphasized in this book, and seen in experiments: A fractal looks the same whatever the scale. Look at some nice cumulus clouds, for example. They consist of big heaps with smaller bulges that have smaller bumps with bumps on them and so on down to the smallest scale you can resolve. In fact, from a picture showing only the clouds one cannot estimate the size of the clouds without extra information.

The fractals we discuss may be considered to be sets of points embedded in space. For example, the set of points that make up a line in ordinary Euclidean space has the topological dimension $D_T = 1$, and the Hausdorff-Besicovitch dimension $D = 1$. The Euclidean dimension of space is $E = 3$. Since $D = D_T$ for the line it is not fractal according to Mandelbrot's definition, which is reassuring. Similarly the set of points that form a surface in $E = 3$ space has the topological dimension $D_T = 2$, and $D = 2$. Again an ordinary surface is not fractal independent of how complicated it is. Finally a ball or sphere has $D = 3$ and $D_T = 3$. These examples really serve to define some of the types of sets of points we discuss.

---

[1]Mandelbrot, private communication, 1987.

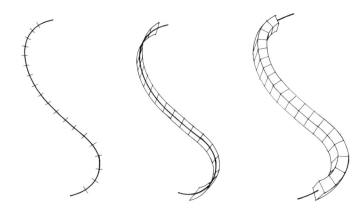

Figure 2.5: Measuring the 'size' of curves.

The concept of a distance between points in space is central to the definition of the Hausdorff-Besicovitch dimension and therefore of the fractal dimension D. How do we measure the 'size' of a set S of points in space? A simple way to measure the length of curves, the area of surfaces or the volume of an object is to divide space into small cubes of side δ as illustrated in figure 2.5. We might use small spheres of diameter δ instead. If we center a small sphere on a point in the set then all points that are at a distance $r < \frac{1}{2}\delta$ from the point at the center are covered by the sphere. By counting the number of spheres needed to cover the set of points we obtain a measure of the size of the set. A curve can be measured by finding the number $N(\delta)$ of line segments of length δ needed to cover the line. For an ordinary curve we have $N(\delta) = L_0/\delta$, of course. The length of the curve is given by

$$L = N(\delta)\,\delta \xrightarrow[\delta \to 0]{} L_0 \delta^0.$$

In the limit $\delta \to 0$, the measure L becomes asymptotically equal to the length of the curve and is independent of δ.

We may choose to associate an *area* with the set of points defining a curve by giving the number of disks or squares needed to cover the curve. This number of squares is again $N(\delta)$, and each square has an area of δ^2. The associated area is therefore given by

$$A = N(\delta)\,\delta^2 \xrightarrow[\delta \to 0]{} L_0 \delta^1.$$

Similarly we may associate a *volume*, V, with the line as follows:

$$V = N(\delta)\,\delta^3 \xrightarrow[\delta \to 0]{} L_0 \delta^2.$$

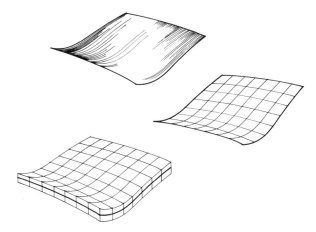

FIGURE 2.6: Measuring the 'size' of a surface.

Now, for ordinary curves both A and V tend to zero as δ vanishes, and the only interesting measure is the length of the curve.

Consider next a set of points that define a surface as shown in figure 2.6. The normal measure is the area A, and we have

$$A = N(\delta)\delta^2 \xrightarrow[\delta \to 0]{} A_0\delta^0.$$

Here one will find that for an ordinary surface the number of squares needed to tile it is $N(\delta) = A_0/\delta^2$ in the limit of vanishing δ, where A_0 is the area of the surface.

We may associate a *volume* with the surface by forming the sum of the volumes of the cubes needed to cover the surface:

$$V = N(\delta)\delta^3 \xrightarrow[\delta \to 0]{} A_0\delta^1.$$

As expected, this volume vanishes as $\delta \to 0$.

Now, may we associate a *length* with a surface? Formally we may simply take the measure

$$L = N(\delta)\delta \xrightarrow[\delta \to 0]{} A_0\delta^{-1},$$

which diverges for $\delta \to 0$. This result is reasonable since it is impossible to cover a surface with a finite number of line segments. We conclude that the only useful measure of a set of points defined by a surface in three-dimensional space is the area.

We shall see that one may easily define sets of points that are curves which twist so badly that their length is infinite, and in fact there are curves (Peano curves) that fill the plane. Also there are surfaces that fold so wildly that they fill space. In order to discuss such strange sets of points it is useful to generalize the measure of size just discussed.

So far, in order to give a measure of the size of a set of points, S, in space we have taken a test function $h(\delta) = \gamma(d)\delta^d$ — a line, square, disk, ball or cube — and have covered the set to form the *measure* $M_d = \sum h(\delta)$. For lines, squares and cubes the geometrical factor $\gamma(d) = 1$. We have $\gamma = \pi/4$ for disks, and $\gamma = \pi/6$ for spheres. In general we find that, as $\delta \to 0$, the measure M_d is either zero or infinite depending on the choice of d — the *dimension* of the measure. The Hausdorff-Besicovitch dimension D of the set S is the *critical dimension* for which the measure M_d changes from zero to infinity:

$$M_d = \sum \gamma(d)\delta^d = \gamma(d)N(\delta)\delta^d \xrightarrow[\delta \to 0]{} \begin{cases} 0, & d > D \; ; \\ \infty, & d < D \; . \end{cases} \qquad (2.3)$$

We call M_d the d-measure of the set. The value of M_d for $d = D$ is often finite but may be zero or infinite; it is the position of the jump in M_d as a function of d that is important. Note that this definition defines the Hausdorff-Besicovitch dimension D as a *local* property in the sense that it measures properties of sets of points in the limit of a vanishing diameter or size δ of the test function used to cover the set. It also follows that the fractal dimension D may depend on position. Actually there are several fine points that have to be considered. In particular, the definition of the Hausdorff-Besicovitch dimension allows for a covering of the set by 'balls' that are not all the same size, but have diameters less than δ. The d-measure is then the *infimum*: roughly the minimal value obtainable in all possible coverings. See section 5.2 for examples. A mathematical treatment is found in the book by Falconer (1985).

The familiar cases are $D = 1$ for lines, $D = 2$ for planes and surfaces, and of course $D = 3$ for spheres and other finite volumes. As we shall see in numerous examples below there are sets for which the Hausdorff-Besicovitch dimension is noninteger and is said to be fractal.

The definition (2.3) of the fractal dimension can be used in practice. Consider again the coastline shown in figure 2.1, which we have covered with a set of squares with edge length δ, with the unit of length taken to equal the edge of the frame. Counting the number of squares needed to cover the coastline gives the number $N(\delta)$. Now we may proceed as implied by equation (2.3) and calculate $M_d(\delta)$, or we may simply go ahead and find $N(\delta)$ for smaller values of δ. Since it follows from equation (2.3), that asymptotically in the limit of small δ

$$N(\delta) \sim \frac{1}{\delta^D} \; , \qquad (2.4)$$

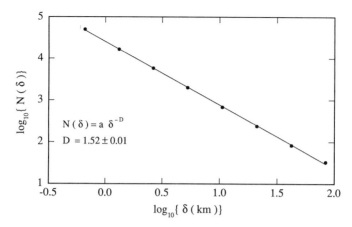

FIGURE 2.7: The number of 'boxes' of size δ needed to cover the coastline in figure 2.1 as a function of δ. The straight line in the log–log plot is a fit of $N(\delta) = a\delta^{-D}$ to the observations. The fractal dimension $D \simeq 1.52$.

we may determine the fractal dimension of the coastline by finding the slope of $\ln N(\delta)$ plotted as function of $\ln \delta$. The resulting plot for the coastline shown in figure 2.1 is presented in figure 2.7. We find approximately $D \simeq 1.5$. The dimension D, determined from equation (2.4) by counting the number of boxes needed to cover the set as a function of the box size, is now called the *box counting dimension* or *box dimension*.

2.4   The Triadic Koch Curve

Figure 2.8 shows the construction of the triadic Koch curve. The triadic Koch curve is one of the standard examples used to illustrate that a curve may have a fractal dimension $D > 1$.

The construction of the Koch curve starts with a line segment of unit length $L(1) = 1$. This starting form is called the *initiator* and may be replaced by a polygon such as an equilateral triangle, a square or some other polygon. The initiator is the 0-th generation of the Koch curve. The construction of the Koch curve proceeds by replacing each segment of the initiator by the *generator* shown as the curve marked $n = 1$ in figure 2.8. Thus we obtain the first generation, which is a curve of 4 line segments each of length $1/3$. The length of the curve is now $L(1/3) = 4/3$. The next generation is obtained by replacing each line segment by a scaled-down version of the generator. Thus in the second generation we have a curve consisting of $N = 4^2 = 16$ segments each having length $\delta = 3^{-2} = 1/9$, and

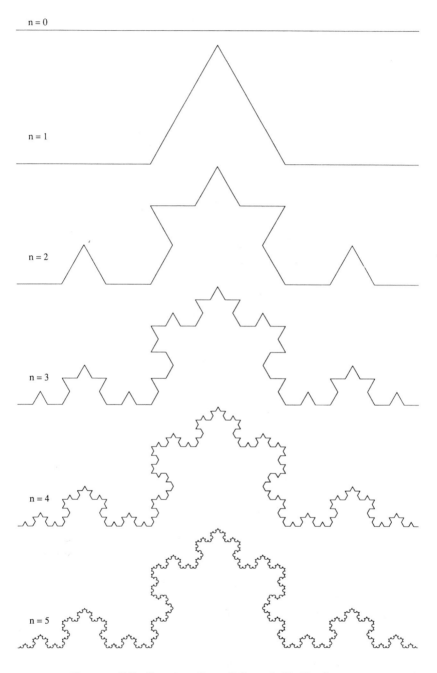

FIGURE 2.8: Construction of the triadic Koch curve.

the length of this curve is $L(1/9) = (4/3)^2 = 16/9$. By applying a reduced generator to all segments of a generation of the curve a new generation is obtained. Such a curve is called a *prefractal*.

For once, let us follow the details of how the expression for D is obtained. The length of the n-th generation prefractal is given by

$$L(\delta) = (4/3)^n .$$

The length of each of the small line segments is

$$\delta = 3^{-n} .$$

By noting that the generation number n may be written in the form

$$n = -\ln \delta / \ln 3 ,$$

we find that the length may be expressed as follows:

$$L(\delta) = (4/3)^n = \exp\left(-\frac{\ln \delta \, [\ln 4 - \ln 3]}{\ln 3}\right) = \delta^{1-D} . \qquad (2.5)$$

This result has the form of equation (2.1) with

$$D = \ln 4 / \ln 3 \sim 1.2628 .$$

The number of segments is $N(\delta) = 4^n = 4^{-\ln \delta / \ln 3}$ and may be written in the form

$$N(\delta) = \delta^{-D} . \qquad (2.6)$$

We shall see that D is the fractal dimension of the triadic Koch curve. First we note that the length of any generation of the Koch construction is a normal curve with a finite length. Mandelbrot calls such curves *prefractals*. However, as we let the number of generations increase δ tends to zero and the length of the curve diverges. Clearly the set of points defined in the limit of an infinite number of iterations of the Koch prescription is not a curve for which length is a useful measure. If, however, we choose to use the test-function $h(\delta) = \delta^d$, we find the d-measure

$$M_d = \sum h(\delta) = N(\delta) h(\delta) = \delta^{-D} \delta^d .$$

We see that the measure M_d remains finite and equal to 1 only when the dimension d of the test function $h(\delta)$ equals D. We conclude that the *critical dimension* and therefore the *Hausdorff-Besicovitch dimension* for the triadic Koch curve is given by $D = \ln 4 / \ln 3$. Now, at each stage of the construction the Koch prefractals may be stretched to form a straight *line*, and therefore we conclude that the topological dimension of the triadic Koch curve is $D_T = 1$. Since the Hausdorff-Besicovitch dimension D for the Koch curve exceeds its topological dimension D_T, we conclude that the Koch curve is a *fractal* set with the fractal dimension $D = \ln 4 / \ln 3$.

2.5   Similarity and Scaling

A line is a special set of points in space. If we change the length scale, we recover the same set of points. In addition we may translate the points of the set and recover the same set of points. The line is *invariant* with respect to translation and scaling — we say that the line is *self-similar*.

To be more precise, let us specify points in space by giving the Cartesian coordinates $\mathbf{x} = (x_1, x_2, x_3)$. A line through the point $\mathbf{x}_0$, in a direction given by $\mathbf{a} = (a_1, a_2, a_3)$, is the set $\mathcal{S}$ of points determined by

$$\mathbf{x} = \mathbf{x}_0 + t\,\mathbf{a}, \quad -\infty < t < \infty.$$

Here the parameter t is any real number. If we change the length scale by the *same* factor r for all of the components of $\mathbf{x}$, then points $\mathbf{x}$ map into new points $\mathbf{x}' = r\mathbf{x} = (rx_1, rx_2, rx_3)$ and we obtain a new set of points $r(\mathcal{S})$, given by

$$\begin{aligned} \mathbf{x}' &= r\,(\mathbf{x}_0 + t\,\mathbf{a}) \\ &= \mathbf{x}_0 + t'\mathbf{a} - (1 - r)\,\mathbf{x}_0. \end{aligned} \tag{2.7}$$

Here $t' = rt$ is again any real number. If we move the new set of points, $r(\mathcal{S})$, by translating all the points an amount $(1 - r)\,\mathbf{x}_0$, the original set of points, $\mathcal{S}$, is recovered — the line is invariant under change of length scale. Also the line is invariant under the translations $\mathbf{x} \to \mathbf{x} + \mathbf{a} \cdot n$, with n any real number.

Using the same type of arguments one concludes that a plane is invariant under translations in the plane, and under change of length scales. Finally the three-dimensional space is invariant under translations in any direction, and under change of length scales.

Other sets of points cannot have these strong symmetries of translational and scaling invariance. A circle is not invariant under translation or under a change of scale — but it is invariant under rotations around its center. Fractals *must* give up some or all of these simple invariances.

It is useful to consider *bounded* sets such as a finite piece of a line. A finite line segment does not have translational symmetry — moving it always results in a new set of points. However, if we change lengths by the scale factor r less than one, we generate a new set of points $\mathcal{S}' = r(\mathcal{S})$ that is a small piece of a line. This line segment may be translated to cover a part of the original line $\mathcal{S}$. If we have chosen r properly, we may cover the original line once with N nonoverlapping segments. We say that the set $\mathcal{S}$ is *self-similar* with respect to the scaling ratio r. For a line segment of unit length we may choose $r(N) = 1/N$, with N any integer. A rectangular piece of a plane may be covered by scaled-down versions if we change the length scales by $r(N) = (1/N)^{1/2}$. Similarly, a rectangular parallelepiped

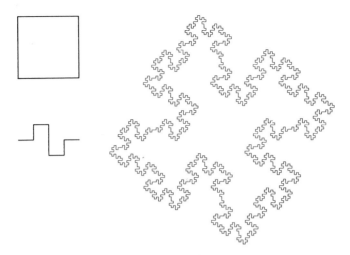

FIGURE 2.9: Construction of the quadratic Koch curve.

may be covered by scaled versions if we use $r(N) = 1/N^{1/3}$. In general we use a scale factor given by

$$r(N) = (1/N)^{1/d} . \tag{2.8}$$

The *similarity dimension, d,* is 1, 2 and 3 for lines, planes and cubes respectively.

Now, consider the Koch curve in figure 2.8. With a scale factor $r = 1/3$, we obtain the first third of the whole curve. We need $N = 4$, such pieces to cover the original set by repeated translations and rotations of this scaled-down piece. We can also scale with a factor $r = (1/3)^n$, using $N = 4^n$ pieces to cover the original set. For the triadic Koch curve we see that the scale factor $r(N)$ satisfies

$$r(N) = (1/N)^{1/D}, \tag{2.9}$$

with the similarity dimension d, equal to the Hausdorff-Besicovitch dimension $D = \ln 4/\ln 3$.

In general we define the *similarity dimension* D_S from the equation above as

$$D_S = -\ln N/\ln r(N). \tag{2.10}$$

The Hausdorff-Besicovitch dimension D equals D_S for self-similar fractals — and we drop the index S for such fractals.

This similarity dimension is easy to determine for various fractals obtained by variants of the Koch construction. Consider a Koch prefractal

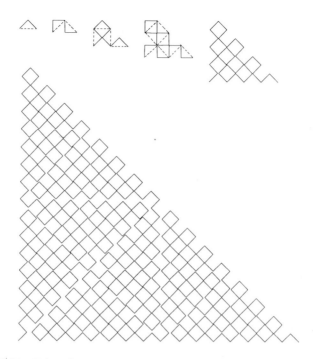

FIGURE 2.10: Triangle sweep with $D = 1.944$. For the first few generations we also show the previous generation as dashed lines. Each of the generations shown is magnified so that the structure of the curve becomes apparent.

constructed starting with the unit square as the initiator and the generator consisting of $N = 8$ pieces each of length $r = 1/4$, as shown in figure 2.9. This curve has the similarity dimension $D = -\ln 8/\ln 1/4 = 3/2$, and again it equals the Hausdorff-Besicovitch dimension of the set obtained after an infinite number of iterations. Note, however, that since we use the unit square as the initiator we find that the figure as a whole is *not* scaling. Each piece of the 'coastline' is self-similar, but when we scale the whole figure by r, we find a smaller version of the original, and the original may *not* be covered using these smaller sets. The point is that fractal scaling behavior is obtained only in the limit $\delta \to 0$, and we conclude that the fractal nature of Koch curves is strictly speaking a *local* property. The Koch construction shown in figure 2.10 is interesting. The curve does not intersect itself but may actually be made to fill a right isosceles triangle. The initiator is the unit interval, and the generator shown in figure 2.10 consists of $N = 2$ pieces of length $r = 0.99 \cdot 1/\sqrt{2}$. We have chosen the

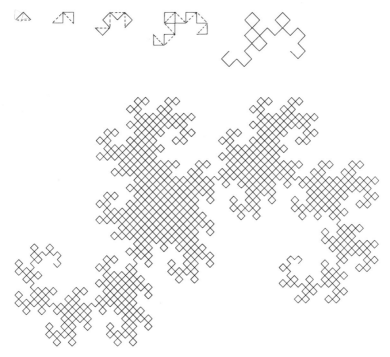

FIGURE 2.11: The Harter-Heightway dragon. $D = 2$.

factor 0.99 to make it easier to see the structure of the curve, since with $r = 1/\sqrt{2}$, each generation is simply a piece of square graph paper.

The final fractal set defined by this construction has the dimension $D = -\ln 2/\ln(0.99/\sqrt{2}) = 1.944$. As seen in the figure, the generator is actually applied in two versions: one that displaces the line segment midpoint to the left, and one that displaces the midpoint to the right. Also each new generation of prefractals starts with alternating left and right generators. In the figure each new generation is magnified to make the segments of a given length so the structure of the curve can be seen without loss of resolution.

Now let us change the rules of construction slightly: Let the first application of the generator displace the midpoint of the generating line to the left. Each following generation starts with the generator to the right, and thereafter right and left midpoint displacements alternate as subsequent midpoints are displaced. The first few generations and the 11th generation of the process are shown in figure 2.11. The limiting fractal curve is called the Harter-Heightway dragon.

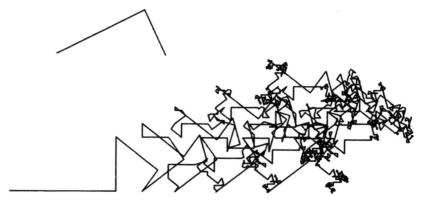

FIGURE 2.12: Modified triangle sweep, $D = 2$. The generator shown in the upper left corner covers the unit interval and scales with two ratios $r_1 = 2/\sqrt{5}$ and $r_2 = 1/\sqrt{5}$.

If we use the generating rule for the triangle sweep but instead use the generator shown in figure 2.12, a self-intersecting plane-filling curve results. The 10th generation of the construction is shown in figure 2.12. The generator breaks the unit interval into two pieces that form an angle of $90°$. The long piece is scaled by the ratio $r_1 = 2/\sqrt{5}$, and the second piece is scaled by a different scale factor $r_2 = 1/\sqrt{5}$. For this case we cannot use equation (2.10) to determine the similarity dimension. Mandelbrot determines the similarity dimension D as the dimension that makes

$$\sum_i r_i^D = 1 \,. \tag{2.11}$$

For the present case we find $D = 2$. Note that it is true, but is not proven here, that this dimension equals the Hausdorff-Besicovitch dimension of the fractal set. Also, in using equation (2.11) we should really discuss how to deal with the overlaps. There are many subtle points to consider once one leaves the simplest fractals.

2.6   Mandelbrot-Given and Sierpinski Curves

The Koch construction shown in figure 2.13 is due to Mandelbrot and Given (1984). The generator for this curve divides the line-segment into pieces of length $r = 1/3$ and adds a *loop* consisting of three pieces; in addition two *branches* are appended.

FIGURE 2.13: An implementation of the Mandelbrot-Given curve. Note that the height of the generator has been reduced slightly so that the structure of the curve becomes apparent. The fractal dimension is $D_B = \ln 8/\ln 3 = 1.89\ldots$ Mandelbrot and Given (1984) also describes random variants of this curve.

Mandelbrot and Given used this and related curves as models of the percolation clusters to be discussed in chapter 7. The curve is interesting in that it has loops of all possible sizes and also branches of all possible lengths. The branches and loops are themselves decorated with loops and branches and so on. In each iteration, from one generation of the prefractal to the next, the generator replaces each line segment in the prefractal by $N = 8$ segments that have been scaled down by the ratio $r = 1/3$. Using the expression (2.10) for the similarity dimension we conclude that the fractal dimension of the Mandelbrot-Given curve is $D = \ln 8/\ln 3 = 1.89\ldots$.

Consider the curve to be made of an electrically conducting material so that a current could be made to flow from the left-hand end to the

FIGURE 2.14: Construction of a Mandelbrot-Given curve without branches. This curve is obtained using a generator with a single loop. The fractal dimension is $D_B = \ln 6 / \ln 3 = 1.63 \ldots$.

right-hand end of the curve. Clearly there would be no flow in any of the branches that result from the two vertical line segments in the generator. The current is confined to flow on the backbone, which is the shape obtained by pruning all the branches that are connected to the original straight line path (the initiator) by only a single bond. If we ignore all the branches then we obtain the curve shown in figure 2.14 (In this implementation the generator is applied in directions that prevent the corners of the loops from touching). The fractal dimension of this curve, which has no 'dangling ends,' is $D_B = \ln 6 / \ln 3 = 1.63 \ldots$ since the generator replaces each line segment by $N = 6$ line segments scaled down by a factor $r = 1/3$. At how many places may we cut a single bond (singly connected bonds) and thereby disconnect the two ends of the initiator? In each application of the generator we generate $N = 2$ singly connected bonds and therefore these bonds form a set of points that has the fractal dimension $D_{sc} = \ln 2 / \ln 3 = 0.63 \ldots$.

The Mandelbrot-Given curves contain many interesting geometrical features, which are not captured by the fractal dimension of the curve as a whole. In fact, such subsets of the curve as the backbone, the singly connected bonds and others are also fractal sets with their own fractal dimensions. It has recently become clear that many physical processes select in a natural way subsets of the structures on which they occur, and

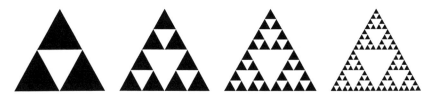

FIGURE 2.15: Construction of the triangular Sierpinski gasket. The initiator is a filled triangle. The generator eliminates a central triangle as shown. The fourth generation of the prefractal is shown to the right. The fractal curve obtained in the limit of an infinite number of generations has the fractal dimension $D = \ln 3/ \ln 2 = 1.58\ldots$.

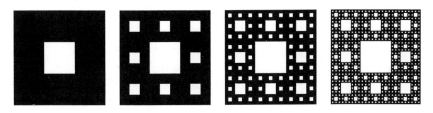

FIGURE 2.16: Construction of the Sierpinski carpet. The initiator is a square and the generator (shown on the left-hand side) is made of $N = 8$ squares. They are obtained by contractions of ratio $r = 1/3$. The right-hand side of the figure shows the fourth construction stage. The similarity dimension is $D = \ln 8/ \ln 3 = 1.89\ldots$.

therefore many fractal dimensions are needed in the discussion of these processes (see chapter 6 for a discussion).

Another construction (see Mandelbrot, 1977, 1982) that creates a curve with loops of all sizes is the Sierpinski gasket shown in figure 2.15. In each application of the generator a filled triangle is replaced by $N = 3$ triangles that have been scaled down by the factor $r = 1/2$ and therefore equation (2.10) shows that the similarity dimension for is $D = \ln 3/ \ln 2 = 1.58\ldots$.   A related curve, the Sierpinski carpet, is shown in figure 2.16. An infinite number of generations of the prefractals leaves a fractal curve. The area (black) visible in the prefractals vanishes in the final fractal curve, while the total perimeter of the holes in the Sierpinski carpet is infinite.

The Sierpinski curves have been used as models for many physical phenomena. Gefen et al. (1980) reported the first systematic study of the critical phenomena that occur near phase transitions in spin systems carried by self-similar fractal lattices. In an interesting experiment Gordon et al. (1986) measured the superconducting-to-normal phase transition temperature $T_c(H)$ as function of the applied magnetic field H of a sample of an aluminum film with the structure of a tenth generation Sierpinski gasket prefractal. The phase-boundary $T_c(H)$ is a self-similar fractal curve and is in excellent quantitative agreement with theoretical predictions.

2.7   More on Scaling

A different point of view is often useful in discussing scale invariance. Consider the Koch curve in figure 2.8 to be the graph of a function $f(t)$. The graph is the set of points (x_1, x_2) in the plane given by the relation $(x_1, x_2) = (t, f(t))$. With a scaling ratio $\lambda = r = (1/3)^n$ for $n = 0, 1, 2 \ldots$, it is clear that the triadic Koch curve has the property

$$f(\lambda t) = \lambda^\alpha f(t),$$

with the scaling exponent $\alpha = 1$. Note that for the Koch curve we have that $f(t)$ is *not* single-valued, but the scaling relation above still holds for any point in the set. The same type of construction may be used on functions defined over all real positive numbers. For example, the power law function $f(t) = bt^\alpha$, satisfies the *homogeneity* relation

$$f(\lambda t) = \lambda^\alpha f(t), \tag{2.12}$$

for *all* positive values of the scale factor λ. Functions that satisfy this relation are said to be *scaling*. Homogeneous functions are very important in the description of the thermodynamics of phase transitions. Much of the progress in recent years in the understanding of critical phenomena near second-order phase transitions can be summarized by stating that such systems have a critical part of their free energy $\mathcal{F}$, which satisfies the scaling form

$$\mathcal{F}_c(\lambda t) = \lambda^{2-\alpha} \mathcal{F}_c(t). \tag{2.13}$$

Here, $t = |T_c - T|/T_c$ is the relative temperature measured from the phase transition temperature T_c, and α is now the specific heat critical exponent. Note that choosing λ so that $\lambda t = 1$ (which is permissible since equation (2.13) is valid for *any* value of λ) gives for the critical part of the free energy the result $\mathcal{F}_c(t) = t^{2-\alpha} \mathcal{F}_c(1)$. Using the thermodynamic definition of the specific heat $\mathcal{C} = -T \partial^2 \mathcal{F}/\partial T^2$, we find that the specific heat behaves for $t \to 0$ as $\mathcal{C} \sim t^{-\alpha}$, consistent with experimental results. Similar scaling behavior describes the statistical properties of *percolation* near the *percolation threshold*; see chapter 7 for a discussion. The modern *renormalization group theory* of critical phenomena explains why the free energy has the scaling form and one may calculate the critical exponents.

Of course, the power-law function and many other functions that exhibit scaling are not fractal curves. However, *scaling fractals* have nice scaling symmetry, and most of the fractals discussed by Mandelbrot are scaling in some sense. He points out that scaling fractals may be used in the description of nature as an approximation — much in the same way we so far have used lines, planes and other smooth surfaces to describe shapes in nature. It is striking how much may be achieved just with scaling fractals, and a thorough study of their properties is certainly rewarding.

2.8   The Weierstrass-Mandelbrot Function

As an example of a scaling fractal curve, we consider the Weierstrass-Mandelbrot fractal function $W(t)$ defined by (Mandelbrot, 1982):

$$W(t) = \sum_{n=-\infty}^{\infty} \frac{(1 - e^{ib^n t})e^{i\phi_n}}{b^{(2-D)n}}. \qquad (2.14)$$

It should be noted that $W(t)$ depends on b in a trivial way since b only determines how much of the curve is visible for a given range of t. The parameter D must be in the range $1 < D < 2$, and ϕ_n is an arbitrary phase — each choice of ϕ_n defines a specific function $W(t)$. This function is continuous, but has no derivative at any point! A simple version is obtained by setting $\phi_n = 0$. The Weierstrass-Mandelbrot cosine fractal function is then the real part:

$$C(t) = \Re\, W(t) = \sum_{n=-\infty}^{\infty} \frac{(1 - \cos b^n t)}{b^{(2-D)n}}. \qquad (2.15)$$

This function has been discussed by Berry and Lewis (1980). The function is believed to be *fractal* with dimension D. It is known to have D as a box dimension, but perhaps not as Hausdorff-Besicovitch dimension. Recently Mauldin (1986) proved that the fractal dimension $D(W_b)$ of the Weierstrass-Mandelbrot function satisfies the following bounds:

$$D - (B/b) \le D(W_b) \le D\,.$$

Here there is a constant B large enough such that the relation is fulfilled for large enough b. We have evaluated the function for a few values of the parameters in the 'time' interval $0 \le t \le 1$, as shown in figure 2.17. The function is reasonably smooth for low values of D, but as D increases toward 2 it fluctuates wildly and one is reminded of noise in electronic circuits. This 'noise' is superimposed on an increasing *trend*. The function $C(t)$ is scaling with a homogeneity equation given by

$$C(bt) = b^{2-D}C(t). \qquad (2.16)$$

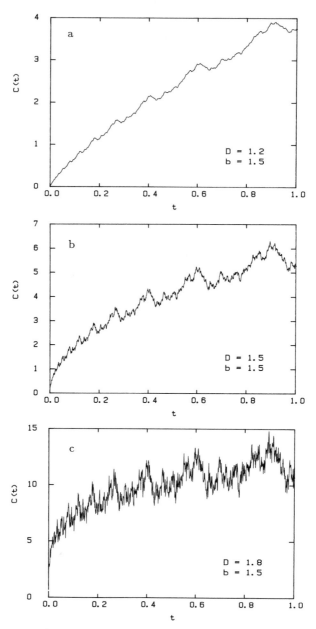

FIGURE 2.17: The Weierstrass-Mandelbrot fractal function $C(t)$ with $b = 1.5$. (a) $D = 1.2$. (b) $D = 1.5$. (c) $D = 1.8$.

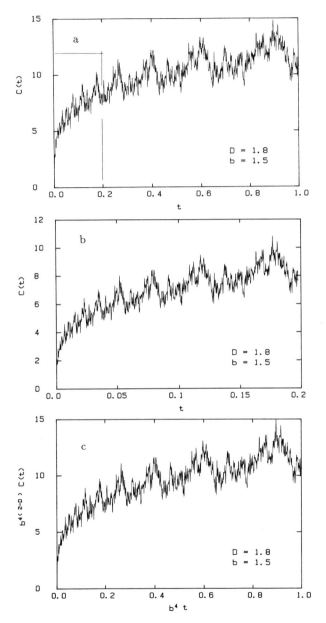

FIGURE 2.18: The Weierstrass-Mandelbrot cosine function with $D = 1.8$ and $b = 1.5$. (a) $0 \leq t \leq 1$. (b) $0 \leq t \leq b^{-4}$. (c) The curve in (b), rescaled to the range 0–1.

Therefore, if we know the function in some range of t, we may deduce the function for any t. As an example compare the function $C(t)$ with $b = 1.5$ and $D = 1.8$, shown in figure 2.18a, to the same function evaluated in the range $0 \leq t \leq b^{-4}$, as shown in figure 2.18b. Clearly the functions look similar. In fact, using equation (2.16), it follows that if we replot the curve in figure 2.18b by replacing t by $b^4 t$ and $C(t)$ by $b^{4(2-D)} C(t)$, as shown in figure 2.18c, the result is the same as the function shown in figure 2.18a. This explicitly exhibits the scaling property of the function $C(t)$.

It is important to note that the graph of the function $C(t)$ is *not* self-similar; it is *self-affine* since we use different scale factors r in the t-direction and in the C-direction. See chapter 10 for a more detailed discussion.

The Weierstrass-Mandelbrot function can be used to generate *random* fractal curves by choosing the phase ϕ_n at random in the interval $(0, 2\pi)$. Some such functions have been discussed by Berry and Lewis (1980). For a recent discussion of the Weierstrass-Mandelbrot function see also Voss (1985a).

Chapter 3

The Cluster Fractal Dimension

The definition of the Hausdorff-Besicovitch dimension D in equation (2.3), and therefore of the fractal dimension for a set of points, requires the diameter δ of the covering sets to vanish. In general, physical systems have a characteristic smallest length scale such as the radius, R_0, of an atom or molecule. In order to apply the ideas of the previous chapter, replace a mathematical line by a linear chain of 'molecules' or *monomers*. As illustrated in figure 3.1, we replace a two-dimensional set of points by a planar collection of monomers, and a volume by a packing of spheres. The number of monomers in a chain of length $L = 2R$ is

$$N = (R/R_0)^1.$$

For a collection of monomers that form a circular disk, we have

$$N = \rho\,(R/R_0)^2.$$

Here the number density is $\rho = \pi/2\sqrt{3}$, for closely packed spheres. For the three-dimensional close packings of spherical monomers into a spherical region of radius R, the number of monomers is given by

$$N = \rho\,(R/R_0)^3,$$

where now the number density is $\rho = \pi/3\sqrt{2}$. These relations apply only in the limit $R/R_0 \gg 1$, because the circular shape of the disk perimeter and the spherical surface of the ball can only approximately be covered by the monomers. For the three cases just discussed we may write the asymptotic form for the relation between the number of particles and the 'cluster' size measured by the smallest sphere of radius R containing the cluster as follows:

$$N = \rho\,(R/R_0)^D, \qquad N \to \infty. \tag{3.1}$$

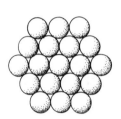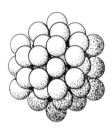

FIGURE 3.1: Simple packings of spheres.

In this *number–radius* relation D is the *cluster dimension*. Since every monomer has the same mass one often interprets N as the mass and ρ as the mass density, and therefore the cluster dimension is often called the *mass dimension*.

The density, ρ, depends on how the monomers are packed. For instance, if spheres are packed at random into a volume, then the density is reduced from $\rho = \pi/3\sqrt{2} = 0.7405$, to 0.637. For other shapes of the clusters ρ includes factors that take into account the *shape* of the cluster. For example, for an ellipsoid of revolution with an axial ratio b/a, we have $\rho = \frac{b}{a}\pi/3\sqrt{2}$, for dense packings of spheres. The cluster dimension, D, on the other hand, does *not* depend on the shape of the cluster, or on whether the packing of monomers is a close packing, a random packing or just a porous packing with a uniform distribution of holes.

It is important to realize that the dimension D, defined by equation (3.1), may be noninteger, i.e., *fractal*. To illustrate this point we return to a discussion of the triadic Koch curve. The construction of the triadic Koch curve in figure 2.8 consists of repeatedly using the *generator* to break up line segments into smaller pieces. A complementary point of view is to consider each prefractal to be a collection of monomers — each monomer representing the generator. In figure 3.2, this method of construction is illustrated. The radius of the monomer, i.e., the generator, is $R_0 = 1/\sqrt{3}$ if the generator spans the unit interval, as usual. The generator itself is the smallest cluster, or the starting generation in a cluster growth process. The first generation contains $N = 4$ monomers and has a radius $R = 3R_0$. In the next generation we have $N = 4^2 = 16$ monomers, and a radius $R = 3^2 R_0 = 9R_0$ for the cluster. In the n-th generation we have

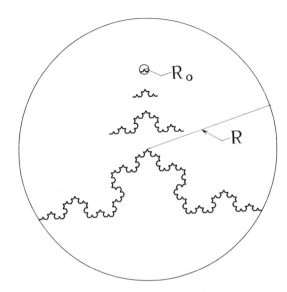

FIGURE 3.2: Triadic Koch clusters.

$N = 4^n$ and $R = 3^n R_0$. It follows that the triadic Koch 'clusters' satisfy the *number–radius relation* in equation (3.1), in the form $N = (R/R_0)^D$, with the cluster dimension equal to the triadic Koch curve fractal dimension $D = \ln 4/\ln 3$. In general, we call the exponent D in the *number–radius* relation the *cluster fractal dimension*.

The cluster fractal dimension is a measure of how the cluster fills the space it occupies. Consider the cluster in figure 3.3 obtained by the *diffusion-limited aggregation process* (DLA). In this process monomers start from far away and diffuse by a random walk process. The wandering monomers stick to the growing cluster when they reach it. This type of aggregation process produces clusters that have a fractal dimension $D(2) = 1.71$, for diffusion in a plane, i.e., for the space dimension $E = 2$. Extensive numeric simulations have been performed,[1] also in $E = 3$-dimensional space, with the result that the clusters are fractal with a dimension $D = 2.50$. For the fractal behavior in 2- to 6-dimensional spaces see Meakin (1983). Recent developments in diffusion-limited aggregation are discussed in the book by Jullien and Botet (1987), and in the review by Meakin (1987b,c).

Let us emphasize that the fact that a cluster is porous or random does not necessarily imply that the cluster is fractal. A fractal cluster has the property that the density decreases as the cluster size increases in a way

---

[1] See Meakin (1983, 1987c), Herrmann (1986), Family and Landau (1984), Stanley and Ostrowsky (1985).

FIGURE 3.3: Cluster resulting from two-dimensional diffusion-limited aggregation. $D = 1.71$. The cluster, containing 50,000 particles, was obtained in an off-lattice simulation where a random walker starts at a random position on a large circle surrounding the cluster. The cluster started at a seed in the center. If the walker contacts the cluster, then it is added to the cluster and another walker is released at a random position on the circle. The cluster was generated using a program developed by Paul Meakin.

described by the *exponent* in the *number–radius relation*. If one insists on introducing a particle density, one finds that the density at a radius r for clusters similar to the one shown in figure 3.3 is given by

$$\rho(r) \sim R_0^{-D} r^{D-E},$$

and this density is constant only if the fractal dimension D equals the Euclidean dimension, E, of the space where the cluster is placed. Fractal clusters have a density that decreases with distance from the origin.

The cluster fractal dimension characterizes a feature of the cluster — how it fills space — in a quantitative way. Note that the *shape* of the cluster is *not* described by the cluster fractal dimension. There are other features

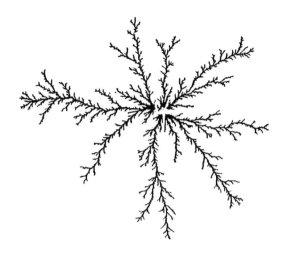

FIGURE 3.4: Metal leaf grown by electrodeposition of zinc at the interface between aqueous zinc sulfate and n-butyl acetate. The fractal dimension is $D \simeq 1.63$ (Matsushita et al., 1984).

of a cluster that may be quantified as well. For instance its *ramification* is a measure of the number of bonds to be cut in order to isolate an arbitrarily large part of the cluster.

The striking patterns generated by the DLA process have been observed in many different types of systems where the growth dynamics is controlled by the Laplace equation. In the next chapter we discuss in some detail the fractal structures generated by the viscous fingering process in Hele-Shaw cells and in porous media. Here we discuss a few of the related structures that have been observed. For example, Niemeyer et al. (1984) observed DLA-like patterns in dielectric breakdown and found a fractal dimension of $D \sim 1.7$. They also introduced the dielectric breakdown model for simulating the process. This model is useful in many contexts and is also related to the DLA model.

Matsushita et al. (1984) observed the DLA-like structure shown in figure 3.4. The zinc metal leaf grows in a two-dimensional manner and has a cluster fractal dimension of $D \simeq 1.63$. Brady and Ball (1984) found that copper electrodeposits, formed under conditions where the growth is diffusion-limited, are three-dimensional fractals with a fractal dimension of $D = 2.43 \pm 0.03$, in agreement with the value of 2.5 for three-dimensional DLA clusters.

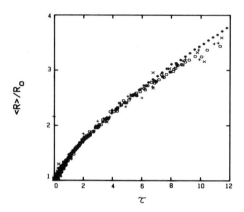

FIGURE 3.5: The effective hydrodynamic radius of IgG aggregates as a function of the reduced time $\tau = \gamma t$, for various temperatures and concentrations. Cluster fractal dimension $D = 2.56 \pm 0.3$ (Feder et al., 1984).

3.1   Measurements of Cluster Fractal Dimensions

Recently measurements of the cluster fractal dimension as given by equation (3.1) have been published for various systems and it has been demonstrated that a description of the experimental results in terms of fractals is useful and leads to a rationalization of the results.

Protein Aggregation

We have studied the aggregation of immunoglobulin proteins of the IgG type using quasielastic light scattering (Feder et al., 1984; Jøssang et al., 1984; Feder and Jøssang, 1985). Immunoglobulins in solution tend to aggregate when heated. The aggregation kinetics is described by the *Smoluchowski equation* (von Smoluchowski, 1916):

$$\frac{d\,n_k}{dt} = \sum_{i+j=k} n_i A_{ij} n_j - 2\sum_j n_k A_{kj} n_j \ . \tag{3.2}$$

Here $n_k(t)$ is the concentration of clusters containing k molecules as a function of time. The probability that a cluster of i molecules combines with another cluster containing j molecules to form a new cluster containing $k = i + j$ molecules is proportional to A_{ij}. We have shown that it follows from equation (3.2) that the effective hydrodynamic cluster radius $\langle R \rangle$, as observed in quasielastic light scattering, grows as a function of time and is given by

$$\langle R \rangle = R_0 \left(1 + \gamma t\right)^{1/D} ,$$

when the clusters satisfy the number–radius relation (3.1), in the form

$$i = (\langle R \rangle / R_0)^{1/D} . \qquad (3.3)$$

Here R_0 is the monomer radius, and γ is a kinetic temperature-dependent constant. From the results shown in figure 3.5 we have concluded that the cluster fractal dimension for IgG aggregates is $D = 2.56 \pm 0.3$.

The results obtained at different temperatures and concentrations all collapse onto a single curve provided the ansatz (3.3) that the clusters are fractal is made. This can be seen in figure 3.5.

Gold Colloid Clusters

Weitz and Oliveria (1984), Weitz and Huang (1984) and Weitz et al. (1985) have studied aggregates of gold colloids by electron microscopy and by light scattering. We have reproduced some of their electron micrographs of gold colloid clusters in figure 3.6. They conclude that these clusters are fractals with a fractal dimension $D \sim 1.75$. The figure shows clearly that the clusters have holes of all sizes that are compatible with the cluster size. Also clusters of different size look similar. The fractal dimension obtained from an analysis of the electron micrographs is $D = 1.7 \pm 0.1$, as shown in figure 3.7. The projections of the clusters seen in the micrographs are not compact, which is consistent with the fact that the observed cluster fractal dimension is less than 2.

Weitz et al. (1985) used both light scattering and small-angle neutron scattering to study the gold colloid aggregates. In both cases the scattering intensity as a function of scattering angle is given by (see also Kjems and Freltoft, 1985)

$$S(q) \sim q^{-D} \quad \text{with} \quad q = \frac{4\pi}{\lambda} \sin \frac{\theta}{2} , \qquad (3.4)$$

where q is the magnitude of the scattering vector, θ is the scattering angle and λ is the wavelength of the radiation used. D is again the cluster fractal dimension as given in equation (3.1) with the modification that R and R_0 are the radii of gyration. From the scattering results they conclude that the fractal dimension of the aggregates is $D \simeq 1.79$, consistent with the results obtained by analyzing the electron micrographs (figure 3.8).

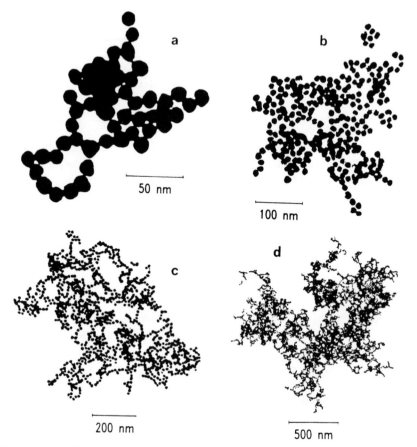

FIGURE 3.6: Transmission electron micrographs of gold clusters of different sizes (Weitz and Huang, 1984).

Silica Colloid Clusters

Schaefer et al. (1984) have studied colloidal aggregates of small silica particles using both X-ray and light scattering. Silica particles with a radius $R_0 \sim 27$ Å were made to aggregate in solution by changes in pH or salt concentration. By the combination of the two different scattering techniques they cover an unusually large range of sizes, as seen in figure 3.9.

Their results cover the range from R_0 up to aggregates with $R \sim 10^4$ Å and their value for the fractal dimension, $D = 2.12 \pm 0.05$, is an unusually precise value. The change in slope of the curve in figure 3.9 is due to the fact

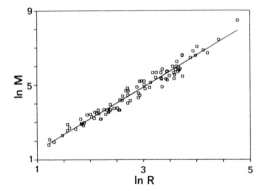

FIGURE 3.7: Variation of the mass as a function of size for gold colloid aggregates. The mass is in units of a single particle mass and the size is scaled by units of the particle diameter. The line corresponds to a fractal dimension $D = 1.7 \pm 0.1$ (Weitz and Huang, 1984).

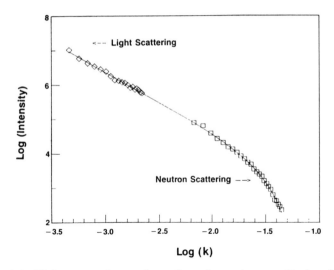

FIGURE 3.8: Light scattering and small-angle neutron scattering from gold colloid clusters, formed by diffusion-limited aggregation, plotted as a function of the scattering wave vector $k = \frac{4\pi}{\lambda} \sin \theta$ (in units of Å^{-1}). The line represents the theoretical fit corresponding to $D \simeq 1.79$ (Weitz et al., 1985).

that the single particles are not fractal clusters, and the intensity crosses over to a behavior characteristic of nonfractal particles when the scattering vector reaches the inverse particle radius a^{-1}.

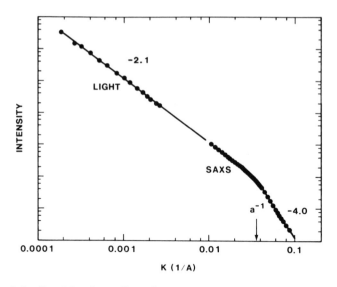

FIGURE 3.9: Combined small-angle X-ray scattering and light scattering intensity results for silica aggregates. $D = 2.12 \pm 0.05$ (Schaefer et al., 1984).

Using small angle-neutron scattering Sinha et al. (1984) have also studied powders of fine silica particles compressed to various densities in the range $0.009\,\text{g/cm}^3$ to $0.45\,\text{g/cm}^3$. For this system they find that the neutron scattering intensity is also given by equation (3.4) and they find the value $D = 2.52 \pm 0.05$ for the fractal dimension of the powders.

Kjems and Freltoft (1985) have used small-angle neutron scattering to study the fractal structure of colloidal silica both in solution and as dry powders. For the solution they find that the colloidal silica is present in the form of clusters with a fractal dimension $D = 2.4 \pm 0.1$, whereas the dry powders at two densities (obtained by compressing the powders) give $D = 2.55 \pm 0.07$.

Silica aerogels are solids that may be prepared as extremely light and tenous materials whose density may be less than $1/_{20}$ of the density of silica. Courtens and Vacher (1987) studied aerogels using elastic coherent small-angle neutron scattering and concluded that the aerogels are fractal with $D = 2.40 \pm 0.03$. They also studied the dynamics of these materials.

Chapter 4

Viscous Fingering in Porous Media

The problem of viscous fingering in porous media is of central importance in oil recovery. It is also an interesting problem in hydrodynamics and in the physics of porous media. It has recently been shown that viscous fingering in porous media is fractal (Måløy et al., 1985a,b; Chen and Wilkinson, 1985). We begin with an introduction to the viscous fingering problem in a two-dimensional geometry, the Hele-Shaw cell, and present some of the relevant experimental results. We then present experimental results on fingering in *porous* media and discuss in particular the very recent evidence that the fingering is *fractal*.

The connection between the diffusion-limited aggregation process — with results similar to figure 3.3 — and the notorious instability of the displacement front in porous media where a high-viscosity fluid (oil) is displaced by a low-viscosity fluid (water or gas) is at first surprising. The analogy that exists between the two phenomena was recently pointed out by Paterson (1984), and rests on the fact that in a continuum approximation both problems are described by *Laplace's* equation.

4.1 Fluid Flow in the Hele-Shaw Cell

A Hele-Shaw cell consists of two transparent plates separated a distance b. Hele-Shaw (1898) studied the flow of water around various objects placed in the cell. He visualized the flow 'streamlines' by injecting a dye to produce colored streamlines. These experiments verify directly that the fluid flow in a Hele-Shaw cell with a small b is the ' potential flow' characteristic for low Reynolds numbers. If the plate separation is increased turbulent flow with confused streamlines arises at moderate flow velocities.

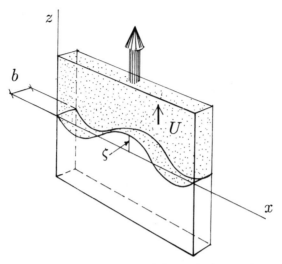

FIGURE 4.1: The geometry of the Hele-Shaw channel.

The equation for the flow velocity $\mathbf{U}$, derived from the Navier-Stokes equations governing flow in the Hele-Shaw cell illustrated in figure 4.1, is

$$\mathbf{U} = -\frac{k}{\mu}\nabla(p + \rho g z) = -M\nabla\phi \,, \qquad (4.1)$$

where p is the pressure, ρ the density and g the component of the acceleration of gravity along the z-coordinate of the cell. The mobility is $M = k/\mu$ and the flow potential is $\phi = (p + \rho g z)$. For a cell placed in the horizontal position we therefore have $g = 0$. The viscosity of the fluid is μ and the *permeability* of the Hele-Shaw cell is

$$k = \frac{b^2}{12} \,. \qquad (4.2)$$

Note that the velocity in equation (4.1) is the *average* velocity over the thickness of the cell. For incompressible fluids the equation of continuity gives

$$\nabla \cdot \mathbf{U} = -\nabla^2(p + \rho g z) = \nabla^2\phi = 0 \,. \qquad (4.3)$$

This is the Laplace equation and it is characteristic of potential problems encountered in electrostatics, in diffusion problems and in many other fields; consequently we call flows controlled by equation (4.3) potential flows. In order to find a solution for the flow velocity we must also specify the boundary conditions — for instance, a given pressure at both ends of the cell, and a vanishing fluid velocity where the fluid is in contact with the walls.

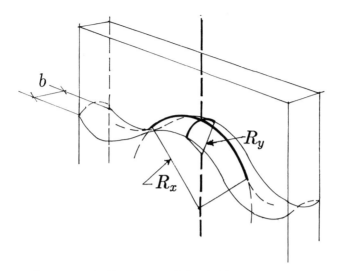

FIGURE 4.2: Geometry of the fluid–fluid interface.

We will discuss the situation illustrated in figure 4.1, where a fluid (index 1) displaces another fluid (2). The interface between the two fluids is controlled by capillary forces when the fluids are at rest and there is a pressure difference between the two fluids

$$(p_1 - p_2) = \sigma \left(\frac{1}{R_x} + \frac{1}{R_y} \right) . \qquad (4.4)$$

Here σ is the interfacial tension between the two fluids. The two principal radii of curvature, R_x and R_y, describe the interface locally as indicated in figure 4.2. We define the radii of curvature to be positive if they have their center in fluid (1). The radius of curvature R_y is controlled by the contact angle, θ, describing how the two-fluid interface contacts the plates that define the cell geometry. Typically one finds that $R_y \sim b/2$, and we will also assume that $R_x \gg R_y$. We have that $p_1 > p_2$ when the fluids are at rest and (2) is the wetting fluid.

Now let us inject the fluid (1) at a constant rate $\mathbf{U}$ at $z = -\infty$ and withdraw fluid (2) at the same rate at $z = \infty$. The interface between the two fluids will then move with a velocity $\mathbf{U} = (0,0,U)$ along the z-axis. However, it turns out that the interface is *unstable* if the viscosity of the driving fluid is smaller than the viscosity of the fluid being driven. Engelberts and Klinkenberg (1951) coined the term *viscous fingering* in relation to their observation of such instabilities when water drives oil out of a porous medium. Flow in porous media also follows equations (4.1) and (4.3), and therefore the flow in Hele-Shaw cells is often used to model the flow in porous media. However, as we shall see, there are important differences and the validity of using the Hele-Shaw cell as a model of flow in porous media is questionable.

The theory of viscous fingering was developed and compared to experiments independently by Saffman and Taylor (1958) and by Chuoke et al. (1959). Recently there has been a growing interest in the field and many new theoretical and experimental results have been published, e.g., those of Bensimon et al. (1986a), Jensen et al. (1987) and DeGregoria and Schwartz (1987). A recent review is given by Homsy (1987).

The physics of viscous fingering lies in the dynamics of the moving boundary. Assume that a pressure difference $\Delta p = p_1(z = 0) - p_2(z = L)$ is maintained over the length L of a finite Hele-Shaw cell where air displaces a high-viscosity fluid. The pressure in the air is constant and equal to the input pressure $p_1(z = 0)$ since we ignore its viscosity. Assume next that a finger gets ahead of the rest of the displacement front; then the pressure at the tip of the finger is also $p_1(z = 0)$. The largest pressure gradient in the high-viscosity fluid is therefore at the tip of the finger and is given by

$$\nabla p = \{p_1(z = 0) - p_2(z = L)\}/(L - z) \,,$$

where z is the position of the tip. This large gradient induces the highest flow velocity in the fluid just ahead of the longest finger which grows faster than the average front — this is clearly an unstable situation.

In order to test the stability of the advancing interface we follow the standard practice (Chuoke et al., 1959; Saffman and Taylor, 1958) and assume that the straight interface is perturbed by a sinusoidal displacement so that in the moving frame of reference the position of the interface is given by the real part of

$$\zeta = \epsilon \exp\left(2\pi\gamma t + i\frac{2\pi x}{\lambda}\right) \,, \tag{4.5}$$

as illustrated in figure 4.1. The wavelength of the perturbation is λ. The growth rate of the perturbation is γ.

For a stable interface the perturbation ζ will decay in time, i.e., $\gamma < 0$. If the growth rate is positive ($\gamma > 0$), a perturbation of infinitesimal amplitude ϵ will grow exponentially.

Solving the equations (4.1)–(4.4), including only terms linear in ζ, gives the result that the front is *unstable* with respect to perturbations that have a wavelength λ that is longer than a *critical wavelength* λ_c given by

$$\lambda > \lambda_c = 2\pi \left(\frac{\sigma}{\left(\frac{\mu_2}{k_2} - \frac{\mu_1}{k_1}\right)(U - U_c)}\right)^{1/2} \,. \tag{4.6}$$

Perturbations with shorter wavelength are stabilized by interfacial tension. Here the critical velocity U_c is given by

$$U_c = \frac{g(\rho_1 - \rho_2)}{\frac{\mu_2}{k_2} - \frac{\mu_1}{k_1}} \ . \tag{4.7}$$

Note that for $\rho_2 > \rho_1$ we find $U_c < 0$ for the case $\mu_2 > \mu_1$ so that the system is unstable even at $U = 0$. Moreover, in the absence of gravity effects $(g = 0)$, the interface is unstable at any velocity since $U_c = 0$.

All wavelengths $\lambda > \lambda_c$ are unstable; however, perturbations with a wavelength λ_m given by

$$\lambda_m = \sqrt{3}\lambda_c$$

have the largest growth rate and will dominate the dynamics of the front. We therefore expect that in experiments in a Hele-Shaw channel of width W an initially straight interface will develop *viscous fingers* with a characteristic period λ_m. Using the expression (4.2) for the permeability and assuming that the viscosity of the driving fluid is negligible $(\mu_1 \ll \mu_2)$, as is the case when glycerol is displaced by air, we find for a horizontal cell that we expect fingers with a period given by

$$\lambda_m = \pi b\sqrt{\frac{\sigma}{U\mu}} = \frac{\pi b}{\sqrt{\text{Ca}}} \ . \tag{4.8}$$

Here we have introduced the dimensionless *capillary number* Ca defined by

$$\text{Ca} = \frac{U\mu}{\sigma} \ , \tag{4.9}$$

which measures the ratio of viscous to capillary forces.

4.2   Viscous Fingers in Hele-Shaw Cells

Saffman and Taylor (1958) and Chuoke et al. (1959) not only developed the theory of viscous fingering in a Hele-Shaw channel, they also studied viscous fingering experimentally. We show fingering patterns for air displacing glycerol observed by Saffman and Taylor in figure 4.3. The initial air–glycerol interface had irregularities at the start of the experiments. Note that the observed wavelength of approximately 2.2 cm is quite close to the wavelength of maximum instability $\lambda_m = \sqrt{3}\lambda_c = 2.1$ cm.

Similar fingers were observed by Chuoke et al., as shown in figure 4.4. Again the period of the finger structure is quite close to the most unstable wavelength λ_m. Recently Maher (1985) observed viscous fingers in a situation where the interface tension could be made exceedingly small.

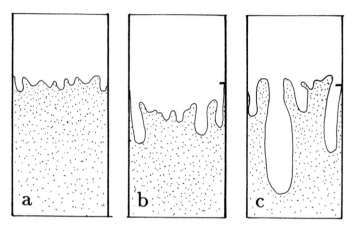

FIGURE 4.3: Viscous fingering in a vertical cell where air displaces glycerol (dark) from the top and downwards. $U = 0.1$ cm/s and $\lambda_c = 1.2$ cm. (a) Early stage with observed average $\lambda \simeq 2.2$ cm. (b) Later stage: Fingers tend to space themselves. (c) Late stage: Longer fingers inhibit the growth of neighbors (Saffman and Taylor, 1958).

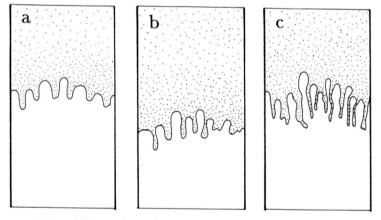

FIGURE 4.4:   Water–glycerol solution (dark), $\mu_1 = 0.552$ poise, $\rho_1 = 1.21$ g/cm^3, displacing oil, $\mu_2 = 1.39$ poise, $\rho_2 = 0.877$ g/cm^3. The system is tilted an angle of $44°25'$. The bulk interfacial tension is $\sigma = 33$ dyne/cm, and the oil wets the walls. The critical velocity is $U_c = 0.23$ cm/s. (a) $U = 0.41$ cm/s, $\lambda_m = 3.5$ cm, observed $\lambda = 3.5$ cm. (b) $U = 0.87$ cm/s, $\lambda_m = 2.6$ cm, observed $\lambda = 2.4$ cm. (c) $U = 1.66$ cm/s, $\lambda_m = 1.6$ cm, observed $\lambda = 1.7$ cm (Chuoke et al., 1959).

If one injects air into the center of a circular Hele-Shaw cell then the growing bubble is unstable with respect to perturbations with a wavelength

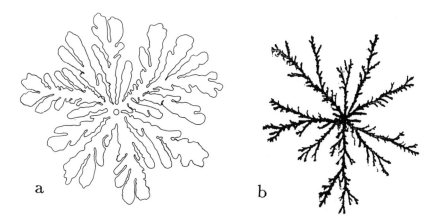

FIGURE 4.5: Radial viscous fingering in a circular Hele-Shaw cell. (a) Air displacing glycerol at Ca = 0.1 (Måløy et al., 1985b). (b) Water displacing a non-Newtonian high-viscosity mixture of scleroglutan in water. This structure is fractal with $D = 1.70 \pm 0.05$ (Daccord et al., 1986).

$\lambda > \lambda_c$ given by equation (4.6). This was shown by Paterson (1981). The wavelength of the perturbation with the highest growth rate again has a wavelength $\lambda_m = \sqrt{3}\lambda_c$. In figure 4.5a we show the fingering pattern obtained at a rather high capillary number Ca. It is interesting to note that as the fingering structure grows outward the fingers widen. However, as soon as they have a width of the order of $2\lambda_m$, the tips split and therefore a branched treelike structure is obtained.

The most unstable wavelength is $\lambda_m \sim \mathrm{Ca}^{-1/2}$, so that it decreases with increasing capillary number. There is a practical limit to how much one may increase the capillary number by increasing the flow velocity U. However, since $\mathrm{Ca} = U\mu/\sigma$, one may increase Ca by using fluids with a small interface tension. Nittmann et al. (1985) used a solution of scleroglu-tan in water as the high-viscosity fluid and displaced it using water. They had a very small interface tension and therefore $\mathrm{Ca} \gg 1$. They observed the fingering structure shown in figure 4.5b. The fingers are now quite narrow and have a width roughly equal to the plate separation. Nittmann et al. measured the fractal dimension of the structure in figure 4.5b using the box counting method and obtained the result $D = 1.7 \pm 0.05$.

Ben-Jacob et al. (1985) made an interesting modification of the Hele-Shaw cell, in which they demonstrated the role of anisotropy. They en-graved a regular sixfold lattice of grooves with depth $b_1 = 0.015$ in., width .0.03 in. and edge-to-edge separation 0.03 in. on the bottom plate of a Hele-Shaw cell 25 cm across. They controlled the effective anisotropy of the cell

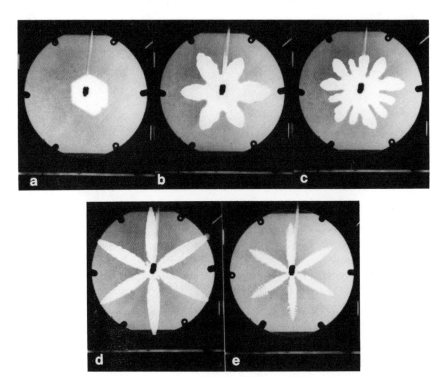

FIGURE 4.6: The various morphologies observed in the Hele-Shaw experiment with anisotropy. The cell is 25 cm across. Glycerol (black) is displaced by air (white) injected at the center. The anisotropy of the cell is
$\alpha = b_1/b_0 = 0.3$. The sequence of figures corresponds to increasing driving
pressure. (a) Faceted growth. (b) Surface tension dendrites (with careful
inspection it is possible to observe that the dendrite tips are pointed at an
angle of 30° to the ruling of the groves). (c) Tip splitting growth (in a
larger and more regular cell they assume that this would correspond to a
dense-branching growth). (d) Kinetic dendrites (the needle crystals grow
parallel to the ruled channels). (e) Kinetic dendrites at higher pressure
(Ben-Jacob et al., 1985).

by varying the plate separation b_0, and defined an anisotropy parameter
by $\alpha = b_1/b_0$.

Nittmann and Stanley (1986) have introduced a modification of the dielectric breakdown model taking into account anisotropy and fluctuations.
Tip splitting in this model is triggered by fluctuations whereas anisotropy
favors dendritic growth. In simulations using this model they were able to

generate patterns that resemble the observed patterns shown in figure 4.6, and in addition they obtained patterns that have a striking resemblance to real snowflakes.

Buka et al. (1986) used an *anisotropic fluid* (a nematic liquid crystal) as the fluid being displaced by air in an ordinary Hele-Shaw cell and observed patterns similar to those shown in figure 4.6b–d. Horváth et al. (1987) found that viscous fingering in a radial Hele-Shaw cell with parallel groves on one of the plates also leads to a rich variety of structures.

4.3   Viscous Fingers in Two-Dimensional Porous Media

The flow of a fluid in a porous medium is controlled by the same equations (4.1) and (4.3) as the flow in a Hele-Shaw cell except that k is now the actual permeability of the medium and not the expression (4.2). However, experiments show that the fingering dynamics is strikingly different. In figure 4.7 we show the results obtained for the displacement of a high viscosity fluid (epoxy) being displaced by air at a moderately high capillary number $Ca = 0.04$ in a two-dimensional porous medium consisting of a randomly packed single layer of glass spheres glued between two glass plates.

The fingering structures observed were analyzed in the following way. The pictures were digitized and each pixel that contained a part of the air (black in the figure) was counted as a 'monomer' and the distance r_i from the center of injection measured. Then in the spirit of the *number–radius* relation (3.1), the number $N(r) = N(r_i < r)$ of pixels containing air inside radius r from the center of injection **was** counted. The total number of pixels containing air is N_0, and the radius of gyration for the fingering structure is $R_g = (N_0^{-1} \sum_i r_i^2)^{1/2}$. In figure 4.8 we have plotted $N(r)/N_0$ as a function of r/R_g, for each of the fingering structures shown in figure 4.7 and for similar experiments using glycerol as a high-viscosity fluid.

It is a remarkable result that the data from all the experiments, i.e., for different fluids and for different times for a given fluid, fall on a single curve. The straight line in figure 4.8 is clear evidence of a number–radius relation of the form

$$N(r) = N_0 \left(\frac{r}{R_g} \right)^D f(r/R_g) . \qquad (4.10)$$

This relation is equation (3.1) modified by a crossover function $f(x)$ that is constant in the range $x < 1$ and tends to x^{-D} for $x > 1$, so that $N(r) \to N_0$ for $r \gg R_g$. The experimental results in figure 4.7 are best

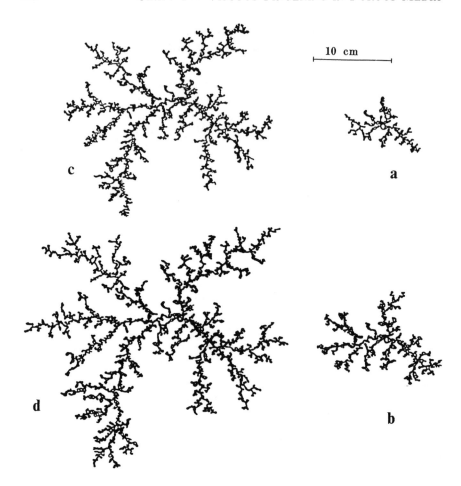

FIGURE 4.7: Fingers of air (black) displacing liquid epoxy in a two-dimensional porous medium consisting of 1.6-mm glass spheres in a monolayer between two glass plates 40 cm in diameter. The center of injection is near the center of the structure. (a) $t = 2$ s after the start of injection. (b) $t = 3.9$ s. (c) $t = 17.2$ s. (d) $t = 19.1$ s. The capillary number is Ca $= 0.04$ (Måløy et al., 1985a,b).

fitted by choosing $D = 1.62 \pm 0.04$, which is the fractal dimension of the viscous fingering in two-dimensional porous media.

Why are the experimental results in the Hele-Shaw cell at comparable capillary numbers shown in figure 4.5a and in figure 4.7 so strikingly dif-

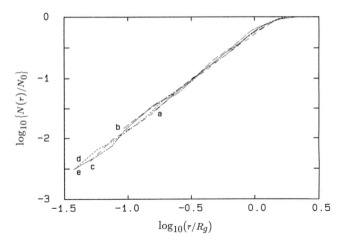

FIGURE 4.8: The normalized finger structure volume $N(r)/N_0$ as a function of the reduced radius (r/R_g) for the structures shown in the previous figure. (a) $R_g = 1.7$ cm. (b) $R_g = 2.9$ cm. (c) $R_g = 5.2$ cm. (d) $R_g = 6.6$ cm. (e) Another experiment with air displacing glycerol at Ca = 0.15, in a 40-cm disk of 1-mm spheres and $R_g = 6.7$ cm (Måløy et al., 1985a,b).

ferent? In both cases we may neglect the viscosity of the driving fluid, and the flow dynamics of the high-viscosity flow is controlled by the Laplace equation. The important difference between the Hele-Shaw cell and the cell containing a porous medium lies in the boundary conditions. In the Hele-Shaw cell the plate separation b is the only length scale in the problem apart from the diameter of the circular cell. In a two-dimensional porous medium, i.e., a circular cell with a *single* random layer of glass spheres, the typical pore size is also equal to b, and therefore the fluid flow problem is controlled by the 'microscopic' length scale in all space directions. Therefore, although the *average* fluid flow velocity U is indeed given by equation (4.1), which is now called *Darcy's equation*, and the incompressibility condition (4.3) also applies, so that a Laplace equation results for the pressure $p(\mathbf{r})$, the problems are actually quite different since the boundary conditions are different. In the Hele-Shaw cell the length scale in the plane of the cell is set by capillary forces, in fact by the critical wavelength λ_c, whereas in the porous medium the length scale is always set by the pore size.

The dynamics of the front is therefore entirely different. For the Hele-Shaw cell it is only a matter of pressure distribution and of satisfying the boundary conditions at the two plates. In the porous medium the pressure is also everywhere determined by the Laplace equation (4.3). However, the

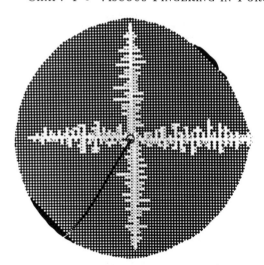

FIGURE 4.9: Viscous fingering of air displacing glycerol in a two-dimensional porous medium consisting of a regular square array of cylinders 1 mm in diameter between two plastic plates. The capillary number is Ca = 0.05 (Feder et al., 1986).

decision to displace the fluid from a given pore at the interface is made not of the basis on the absolute value of the pressure difference between the air and the fluid, but rather on the value of the pressure relative to the capillary pressure associated with the pore neck leading to that pore, since it is more difficult for the air to enter a narrow pore. This last step introduces randomness into the problem since the width of the pore necks is random with some size distribution. The dynamics of the viscous fingering front in porous media therefore has two main components: the global pressure distribution controlled by Darcy's law and therefore the Laplace equation, and the local fluctuations in pore geometry. The result of these two factors is a growing *fractal* structure.

That randomness on the pore level is required in order to obtain fractal fingering may be seen in figure 4.9, where the regular fingering structure clearly is not fractal. Indeed, Chen and Wilkinson (1985) have shown, by experiments and simulations, that randomness in the pore structure is a requirement for fractal fingering.

If instead of displacing the fluid at very high capillary numbers, one performs displacements at very *low* capillary numbers, Ca $\sim 10^{-4}$, then one observes fractal structures characteristic of *invasion percolation* (see section 7.8).

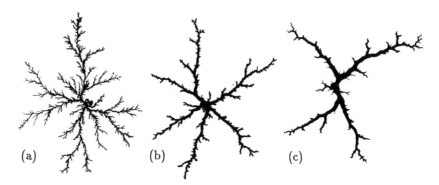

FIGURE 4.10: Photographs of two-dimensional patterns resulting from chemical dissolution of porous plaster by the radial flow of water at different injection rates: (a) 48 cm^3/h; (b) 4 cm^3/h; (c) 2 cm^3/h (Daccord, 1987).

Chemical Dissolution of a Porous Medium

Daccord (1987) has studied the chemical dissolution of a porous medium by a reactive fluid that flows through the medium. Daccord cast a 1-mm-thick plate of pure plaster ($CaSO_4 \cdot 0.5\,H_2O$) between two transparent plates. He injected distilled water at the center of the disk, thereby creating a flow in the porous plaster. Plaster is slightly soluble in distilled water, and the flowing water therefore gradually etched a pattern in the plaster. Daccord made a cast of the etched pattern using Woods metal, and dissolved away the remaining plaster. The resulting patterns for several flow rates are shown in figure 4.10.

The dissolution patterns found at high flow rates are strikingly similar to the viscous fingering patterns observed in two-dimensional porous flow (figure 4.7) and in DLA simulations (figure 3.3). Daccord has introduced a modification of the DLA model that takes into account the transport of dissolved material. Simulations with this model reproduce the observed patterns very nicely. The high flow rate patterns are essentially DLA patterns and have a fractal dimension of $D = 1.6 \pm 0.1$. Daccord and Lenormand (1987) have used this technique to study three-dimensional dissolution by flow in porous media.

4.4   Viscous Fingering and DLA

The diffusion-limited aggregation process represents a problem in which particles are left to wander at random until they reach the 'surface' of the

Withdrawal

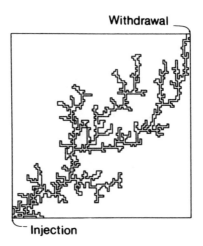

Injection

FIGURE 4.11: Simulation of the displacement front at breakthrough for one quarter of a five-spot well pattern. The driving fluid has infinite mobility (Paterson, 1984).

cluster where they come to rest and thus make the surface grow one step at the point of attachment. In the continuum limit random walkers are described by the diffusion equation. Let $C(\mathbf{r}, t)$ be the concentration of walking particles; then the diffusion equation may be written

$$\frac{\partial C(\mathbf{r}, t)}{\partial t} = \mathcal{D}\nabla^2 C(\mathbf{r}, t) . \tag{4.11}$$

The diffusion constant is in general given by the *Einstein relation*, $\mathcal{D} = \frac{1}{2}\Gamma a^2$, for particles that take steps of length a in random directions at a rate Γ.

With a steady supply of walkers a steady state, $\partial C/\partial t = 0$, may be obtained and for this case the diffusion equation (4.11) reduces to the Laplace equation

$$\nabla^2 C = 0. \tag{4.12}$$

A locally smooth boundary moves with a velocity given by

$$V_\perp = -\mathcal{D}\,\mathbf{n} \cdot \nabla C \mid_s , \tag{4.13}$$

in the direction perpendicular to the surface with surface normal $\mathbf{n}$ (Witten and Sander, 1983).

The flow of a fluid in a porous medium is described by *Darcy's equation* (4.1). For incompressible flow it is reduced to the Laplace equation (4.3). This is the same equation as (4.12) above with ϕ replacing the concentration.

However, the problem of two-fluid displacement in porous media is more complicated. The standard approach is to take equation (4.1) to be valid in each of the fluid components. It has been known for a long time that the displacement front is unstable and forms 'fingers' if the viscosity of the driving fluid μ_1 is lower than the viscosity μ_2 of the fluid being driven. In the limit that the viscosity of the driving fluid can be neglected — for instance in a gas drive — the displacement front moves with velocity $M\nabla\phi$, analogous to equation (4.13) for DLA. Paterson (1984) first pointed out the analogy between DLA and flow in porous media. He has made simulations for the standard five-spot well pattern, in which the driving fluid is injected in a central well and fluids are produced at four other wells at the corners of a square. A quarter of a square is shown in figure 4.11. To the extent that the analogy holds, the fractal dimension of the displacement front should be 1.7, as for DLA in the plane, or 2.5, as for three-dimensional displacement. For finite mobility ratios the discussion must be modified, however, since in the opposite limit, where the driving fluid is more viscous, the front is known to be stable and therefore must have a fractal dimension of 1 in the two-dimensional case, and 2 in the three-dimensional displacement.

We have seen earlier in this chapter that DLA and viscous fingering in porous media look very much the same. Also the structures formed have roughly the same fractal dimension. We may illustrate the effect of changing the boundary condition for the flow controlled by the Laplace equation. In figure 4.12 we show the result of a DLA simulation in which diffusing particles are released from the top line. If a particle contacts the bottom line it becomes the root of a new tree. A particle that contacts any one of the existing trees becomes a part of the tree at the point of contact. A new particle is started on its random walk trajectory at a random position on the top line as soon as the previous one has been adsorbed. The diffusing particles are reflected at the vertical walls. The results of the simulations and the experiments in figure 4.13 clearly look quite similar. In models that are quasi two-dimensional, consisting of many layers of particles but otherwise in a Hele-Shaw channel geometry, earlier experiments by Paterson et al. (1982) and by Lenormand and Zarcone (1985b) also show treelike viscous fingering structures at high capillary numbers. Matsushita et al. (1985) have observed similar fractal tree structures in electrolytically grown zinc metal.

In a recent modification of the DLA model, one may also obtain the dynamics of the process. Måløy et al. (1987a,b) and Meakin (1987a,c) show that the modified dynamical DLA process describes in detail the rate at which the fingers grow. We conclude that the DLA simulations describe quite accurately two-dimensional viscous fingering at high capillary numbers.

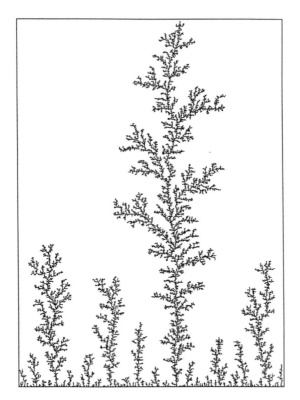

FIGURE 4.12: DLA growth from a line. The particles start on random walk trajectories from the top line and they are reflected from the side walls. The particles attach on contact to the bottom line and to trees connected to the bottom line. The baseline is 801 cells long, and the height of the longest finger is 1099. The number of particles is 47,348 (Hinrichsen et al., 1987).

4.5   Viscous Fingers in Three-Dimensional Porous Media

Several observations of fingering in porous media confirm this picture of fingering qualitatively in three dimensions as well. Engelberts and Klinkenberg (1951) displaced oils with water in packs of sand and sampled various cross sections. The water contained fluorescein and when the cross sections were photographed in ultraviolet light the water fingers became visible as shown in figure 4.14. The rather irregular appearance of the water fingers with cross sections of all sizes begs for a fractal description.

The growth of fingers in porous media in the displacement of oil by water with a viscosity ratio of 80 in transparent models made of compacted

FIGURE 4.13: Viscous fingering in the displacement of glycerol by air (black) in a two-dimensional porous model. The model consists of glass beads 1 mm in diameter arranged as a close-packed random single layer between transparent plates. The model is placed horizontally. Air is injected along a line shown at the bottom in the figure, and the displaced glycerol leaves the model at the line indicated at the top of the figure (Måløy et al., 1987c).

Pyrex powder has been observed by van Meurs (1957) and van Meurs and van der Poel (1958), as shown in figure 4.15.

The general fractal-like structure of the fingers is very similar to the observed structure in DLA clusters.

Another observation of fingering in a five-spot pattern is shown in figure 4.16. Habermann (1960) made consolidated sand packs by coating the grains with a thin layer of epoxy; these were cured between Lucite windows to form models of up to $15 \times 15 \times 1/8$ in. in size. The fluids used in the various experiments were hydrocarbons, alcohols, glycols and water.

There is a striking similarity with Paterson's simulation, aggregation clusters and the fronts observed at high mobility ratios. We believe that Paterson's analogy between aggregation kinetics and displacement fronts in porous media is quite accurate and describes the viscous fingering at

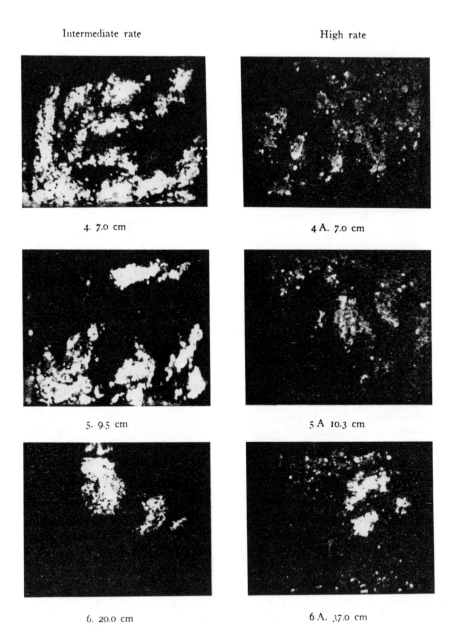

Intermediate rate

High rate

4. 7.0 cm

4 A. 7.0 cm

5. 9.5 cm

5 A 10.3 cm

6. 20.0 cm

6 A. 37.0 cm

FIGURE 4.14: Distribution of water in various cross sections after introduction of 5% of the pore volume. Viscosity ratio $\mu/\mu_0 = 24$ (Engelberts and Klinkenberg, 1951).

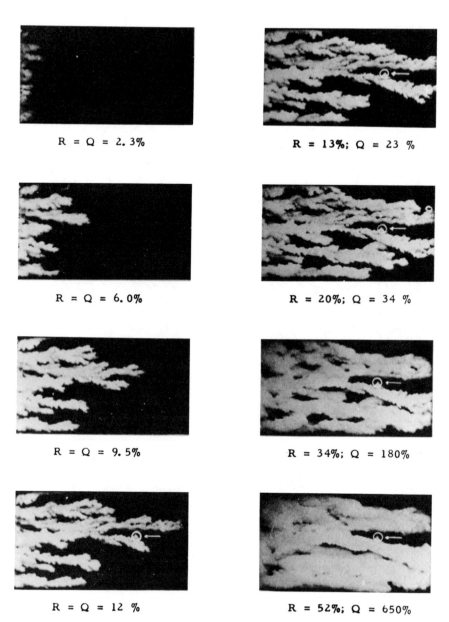

R = Q = 2. 3%   R = 13%; Q = 23 %

R = Q = 6. 0%   R = 20%; Q = 34 %

R = Q = 9. 5%   R = 34%; Q = 180%

R = Q = 12 %   R = 52%; Q = 650%

FIGURE 4.15: Linear displacement of oil by water at oil–water viscosity
ratio 80 in a three-dimensional porous medium. Q—total volume injected,
R—'recovery' volume, expressed as percentages of total pore volume (van
Meurs, 1957; van Meurs and van der Poel, 1958).

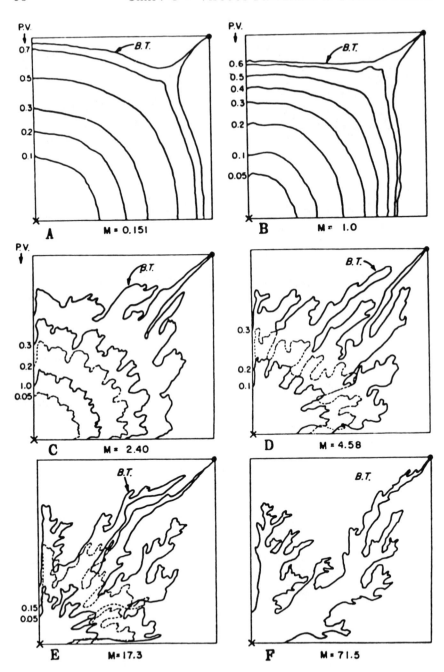

FIGURE 4.16: Displacement fronts for different mobility ratios, M, and injected pore volumes, $P.V.$, until breakthrough, $B.T.$ (Habermann, 1960).

high capillary numbers in two-dimensional media discussed in section 4.3 very well. However, other factors may enter as well. For example, the wetting properties of the fluids and the porous medium also control the finger width in realistic systems. For recent discussions see Lenormand (1985) and Stokes et al. (1986). Also the situation of miscible displacement is different; see, e.g., Chen (1987).

Experiments on three-dimensional systems are much more difficult to analyze quantitatively. We feel, however, that experiments documented in the literature support the picture that the displacement fronts are fractals for high mobility ratios even for three-dimensional displacement. Clément et al. (1985) found evidence of fractal fronts in their three-dimensional experiments. How to extend the analogy to model the situation in which the mobility ratio is finite and/or the capillary number is not large remains an open question.

Chapter 5

Cantor Sets

So far we have introduced several dimensions: the Hausdorff-Besicovitch dimension, the topological dimension, the Euclidean dimension, the similarity dimension, the scaling dimension, the cluster dimension and the box dimension. The Cantor sets illustrate well many of the important and interesting features of fractals.

5.1 The Triadic Cantor Set

A very simple construction due to Cantor generates fractal sets with a fractal dimension in the range $0 < D < 1$. As shown in figure 5.1 the initiator is the unit interval $[0, 1]$, and the generator divides the interval into three equal parts and deletes the open middle part leaving its endpoints. The generator is then applied again to each of the two parts and so on. This procedure very quickly produces extremely short segments. Because of the finite resolution of our graphics we find that already the 6-th generation cannot be distinguished from the 5-th generation. After an

FIGURE 5.1: Construction of the triadic Cantor set. The initiator is the unit interval $[0, 1]$. The generator removes the open middle third. The figure shows the construction of the five first generations. $D = \ln 2 / \ln 3 = 0.6309$.

infinite number of generations what remains is an infinite number of points scattered over the interval. This set is called the *Cantor dust* (Mandelbrot, 1977).

In the following we evaluate the various dimensions introduced in the preceding sections for the Cantor set.

First let us consider the Hausdorff-Besicovitch dimension[1] defined in equation (2.3). In the n-th generation we have $\mathcal{N} = 2^n$ segments each of length $l_i = (1/3)^n$ for $i = 1, \ldots, \mathcal{N}$. If we try to cover the set with line segments of length $\delta = l_i$ and place them carefully we may cover all segments generated in the n-th generation and therefore all points in the Cantor set. The measure defined in equation (2.3) is given by

$$M_d = \sum_{i=1}^{\mathcal{N}} \delta^d = 2^n (1/3)^{nd} = \delta^{d-D} \ .$$

This measure diverges or tends to zero as δ is decreased, unless we choose $d = D = \ln 2/\ln 3 = 0.6309$. The topological dimension of the Cantor set is $D_T = 0$. As $D_T < D$, we conclude that the triadic Cantor set is a fractal set with a fractal dimension given by

$$D = \frac{\ln 2}{\ln 3} \ , \qquad fractal\ dimension. \tag{5.1}$$

The Cantor set as described here is not entirely self-similar. However, we may enlarge the set by an extrapolation procedure that covers the region $[0, 3]$ by two Cantor sets covering the intervals $[0, 1]$ and $[2, 3]$. Repeating this process ad infinitum, we generate a self-similar set on the half-line $[0, \infty)$. Changing the length scale by the factor $r = 1/3$, we need $N = 2$ such pieces to cover the original set. From the definition of the *similarity dimension* D_S in equation (2.10), we have

$$D_S = \frac{\ln N}{\ln(1/r)} = \frac{\ln 2}{\ln 3} \ , \qquad similarity\ dimension. \tag{5.2}$$

The similarity dimension equals the fractal dimension for the triadic Cantor set.

Using the equation (5.2) it is trivial to construct Cantor sets with any given dimension in the range $0 < D < 1$. As an example we show in figure 5.2 two different constructions that both have $D = 1/2$. The two sets 'look' different in spite of the fact that they have the same fractal dimension; they have different *lacunarity* (Mandelbrot, 1982).

---

[1] As actually finding the Hausdorff-Besicovtch dimension is harder — we only present the general idea.

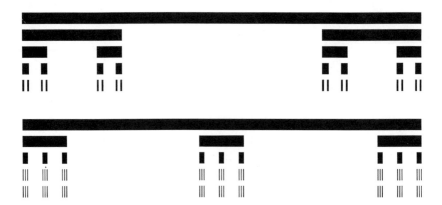

FIGURE 5.2: Two constructions of the Cantor set with $D = 1/2$. Top figure: $N = 2$ and $r = 1/4$. Bottom figure: $N = 3$ and $r = 1/9$.

The *cluster dimension* or *mass dimension* is obtained if we consider the extrapolated version of the Cantor set. Start with 'monomers' of length R_0 and generate a 'cluster' of $N = 2$ monomers of dimension $R = 3R_0$, and so on. A cluster of $N = 2^n$ monomers has a diameter $R = 3^n$, and consequently we find that the cluster fractal dimension defined in equation (3.1) is given by

$$D_C = \frac{\ln N}{\ln R} = \frac{\ln 2}{\ln 3} , \qquad cluster\ dimension\ . \qquad (5.3)$$

The cluster dimension equals the fractal dimension for this Cantor set.

We conclude that the various dimensions defined so far are all equal for the rather simple triadic Cantor set.

5.2   Scaling with Unequal Ratios

What happens when the two segments in the triadic Cantor generator are no longer identical? In figure 5.3 we have drawn the Cantor bar that results when the first section has the length $l_1 = 1/4$ whereas the second part has length $l_2 = 2/5$. Let us evaluate the fractal dimension of this rather simple Cantor set $\mathcal{S}$.

The fractal set $\mathcal{S}$ may be covered by some number $\mathcal{N}$ of disjoint pieces $\mathcal{S}_1, \mathcal{S}_2, \ldots, \mathcal{S}_{\mathcal{N}}$. Let the Euclidean length (diameter) of the i-th set be l_i so that $\mathcal{S}_i$ fits into a (hyper) cube of side l_i. With a partition for which $l_i \leq \delta$ the d-measure used in equation (2.3) to define the Hausdorff-Besicovitch

FIGURE 5.3: A two-scale Cantor bar construction with $l_1 = 1/4$ and $l_2 = 2/5$. The fractal dimension of the Cantor set is $D = 0.6110$.

dimension is

$$M_d = \sum_{i=1}^{\mathcal{N}} l_i^d \xrightarrow[\delta \to 0]{} \begin{cases} 0, & d > D ; \\ \infty, & d < D . \end{cases} \tag{5.4}$$

The critical dimension $d = D$ obtained in the limit $\delta \to 0$ is the fractal dimension of the set. We note that this definition coincides with the definition given by Mandelbrot for Koch curves where different parts scale with the ratios $r_i = l_i$. The *similarity dimension* D_S for such a set is therefore the dimension that satisfies

$$\sum_{i=1}^{N} r_i^{D_S} = 1 , \tag{5.5}$$

as already discussed in connection with equation (2.11).

As an example consider the Cantor set constructed as shown in figure 5.3. In the n-th generation there are $\mathcal{N} = 2^n$ segments. The shortest segment has length $l_1^n = (1/4)^n$ and the longest segment has length $l_2^n = (2/5)^n$. There are in general $\binom{n}{k} = n!/k!(n-k)!$ segments with length $l_1^k l_2^{n-k}$, with $k = 0, 1, \ldots, n$. In the n-th generation the measure M_d is given by

$$M_d = \sum_{i=1}^{\mathcal{N}} l_i^d = \sum_{k=0}^{n} \binom{n}{k} l_1^{kd} l_2^{(n-k)d} = (l_1^d + l_2^d)^n . \tag{5.6}$$

Since n increases to infinity as the length scale $\delta = l_2^n$ tends to zero, we find that M_d remains finite if and only if $d = D$, where D satisfies the equation

$$(l_1^D + l_2^D) = 1 . \tag{5.7}$$

A numerical solution of this equation with $l_1 = 1/4$ and $l_2 = 2/5$ gives $D = 0.6110$.

Chapter 6

Multifractal Measures

Consider a *'population'* consisting of *'members'* distributed over a volume of linear size L, that is, over a volume L^E. The population could, in fact, be the human population distributed over the surface of the earth. The population could also be considered to be the meteorological observation posts, which are unevenly distributed over the globe. The distribution of energy dissipation in space is an example relevant to three-dimensional turbulent flow. The distribution of errors in a transmission line is an example of a one-dimensional population. In physics we routinely consider the distribution of impurities on surfaces and in the bulk. The magnetization of a magnet fluctuates in space. We could consider the local magnetic moments to be members of a population. Many variables fluctuate wildly in space. Gold, for instance, is found in high concentrations at only a few places, in lower concentrations at many places, and in very low concentrations almost everywhere. The point is that this description holds whatever the linear scale is — be it global, on the scale of meters, or on the microscopic scale. *Multifractal measures* are related to the study of a distribution of physical or other quantities on a geometric *support*. The support may be an ordinary plane, the surface of a sphere or a volume, or it could itself be a fractal.

The concepts underlying the recent development of what are now called multifractals were originally introduced by Mandelbrot (1972, 1974) in the discussion of turbulence and expanded by Mandelbrot (1982, p. 375) to many other contexts. The application to turbulence was further developed by Frisch and Parisi (1985) and Benzi et al. (1984). Much of the recent interest in multifractals started out with work by Grassberger (1983), Hentschel and Procaccia (1983b) and Grassberger and Procaccia (1983). A related dimension function was introduced by Badii and Politi (1984, 1985).

The analysis of experimental results and the introduction of the $f(\alpha)$ function by Frisch and Parisi (1985) and Jensen et al. (1985) gave a most remarkable agreement between observations and a simple theoretical model (see section 6.10). They demonstrated the usefulness of multifractals in describing experimental observations. Related work is described by Bensimon et al. (1986b), Halsey et al. (1986b) and Glazier et al. (1986). A recent thermodynamic formulation of multifractals, that of Feigenbaum et al. (1986), maps the measure onto an Ising model. Nonanalyticities in the generalized dimensions of multifractal sets of physical interest may be interpreted as phase transitions, according to Katzen and Procaccia (1987).

The distribution of currents in fractal resistor networks has been discussed in terms that relate directly to multifractals. See for example de Arcangelis et al. (1985), Rammal et al. (1985), Aharony (1986) and Blumenfeld et al. (1987).

Multifractals in the context of diffusion-limited aggregation and related growth processes have been discussed by Meakin et al. (1985, 1986), Meakin (1987b,c) and Halsey et al. (1986a). Nittmann et al. (1987) have analyzed viscous fingering in Hele-Shaw cells and found evidence for multifractal behavior. Måløy et al. (1987b) analyzed the growth dynamics of viscous fingering at high capillary numbers. The *observed growth* measure is described by a multifractal structure. These results are discussed in section 6.12. Fractal aggregates and their fractal measures have recently been reviewed by Meakin (1987c).

The idea that a fractal measure may be represented in terms of intertwined fractal subsets having different scaling exponents opens a new realm for the applications of fractal geometry to physical systems. The study of multifractals is a rapidly developing field. In this chapter we first discuss a few of the basic ideas, illustrating them with simple examples. Later we discuss experimental evidence for multifractal behavior.

6.1   Curdling and the Devil's Staircase

Let us modify the meaning of the Cantor set. The initiator is now no longer considered to be the unit interval, but rather a bar of some material with a density $\rho_0 = 1$. The original bar has a length $l_0 = 1$, and therefore the mass $\mu_0 = 1$. The operation of applying the generator now consists of cutting the bar into two halves of equal mass $\mu_1 = \mu_2 = 0.5$, and then hammering them so that the length of each part becomes $l_1 = 1/3$. By this process the density increases to $\rho_1 = \mu_1/l_1 = 3/2$. Repeating this process, we find that in the n-th generation we have $N = 2^n$ small bars, each with a length

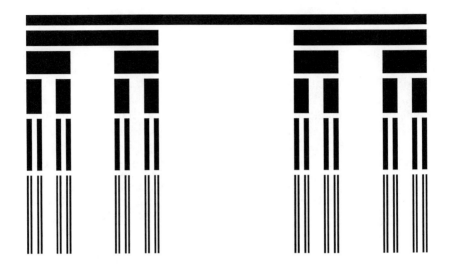

FIGURE 6.1: The triadic Cantor bar. A bar of unit length and mass is
divided into two and hammered to a reduced length so that the density
increases. The height of the bars in the n-th generation is proportional to
the density ρ_i. The Lipschitz-Hölder exponent is $\alpha = (\ln 2)/(\ln 3)$, and the
fractal dimension of the support of the mass is $f = D = (\ln 2)/(\ln 3)$.

$l_i = 3^{-n}$ and a mass $\mu_i = 2^{-n}$ for $i = 1, \ldots, N$. Note that the process
conserves the mass so that

$$\sum_{i=1}^{N} \mu_i = 1 . \tag{6.1}$$

Mandelbrot (1977, 1982) calls this process *curdling* since an originally uniform mass distribution by this process clumps together into many small
regions with a high density.

It follows that the mass of a segment of length l_i where $l_i \leq \delta$ is given
by

$$\mu_i = l_i^\alpha . \tag{6.2}$$

Here the *scaling exponent* α is given by $\alpha = \ln 2/\ln 3$. The density of each
of the small pieces is

$$\rho_i = \frac{\mu_i}{l_i} = \rho_0 \, l_i^{\alpha-1} , \tag{6.3}$$

which diverges as $l_i \to 0$. The scaling exponent α is a classical notion in
mathematics — the Lipschitz-Hölder exponent, which we discuss further
in section 6.4. This exponent controls the singularity of the density and
may also be called the *exponent of the singularity*.

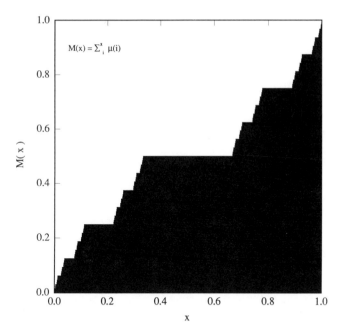

FIGURE 6.2: The mass of the triadic Cantor bar as a function of position along the bar. The curve is called a *Devil's staircase*.

In figure 6.1 we have drawn a version of the triadic Cantor set in which the height of each segment is given by the density ρ_i. We see that this modification of the Cantor construction needs the scaling exponent α to describe how the height of the bars increases as the width decreases. We may say that the singularities with exponent α have a *support* of fractal dimension $f = D$.

In the discussion above we considered μ_i to be the contribution from a segment to the mass in the Cantor bar. The results obtained would be the same if instead of using the mass we consider μ to be electric charge, magnetic moment, hydrodynamic vorticity or a probability for some phenomenon. In general μ can *measure* any quantity that is supported by a geometric set.

An interesting construct — the *Devil's staircase* — may be obtained from a Cantor bar. Start at the left of the bar shown in figure 6.1, at the position $x = 0$, and find the mass contained in the segment $[0, x]$, which formally may be written

$$M(x) = \int_0^x \rho(t)\, dt = \int_0^x d\mu(t) .$$

Here the 'density' $\rho(x)$ is zero in the gaps and infinite on all of the infinite number of points that constitute the Cantor set. The mass $M(x)$ remains constant on the intervals that correspond to the gaps. The lengths of the gaps add up to 1, i.e., the length of the whole interval. Therefore over a length equal to the length of the interval one finds that $M(x)$ does not change. One might then jump to the conclusion that $M(x) \equiv 0$, which would be a correct conclusion for a sensible curve. However, the mass increases, by infinitesimal jumps, at the points of the Cantor set and all these contributions add up to $M(1) = 1$. The mass as a function of x, shown in figure 6.2, resembles a staircase — called the Devil's staircase — that is (almost) everywhere horizontal. The self-affine nature of this curve is apparent in figure 6.2. For a discussion of the origin of Devil's staircases in many physical systems see Bak (1986).

6.2   The Binomial Multiplicative Process

Populations or distributions generated by a multiplicative process have many applications and have the advantage that many properties of these distributions may be easily analyzed. We start with a one-dimensional example. Let a population consisting of $\mathcal{N}$ members be distributed over the line segment $\mathcal{S} = [0,1]$. We will consider the limit $\mathcal{N} \to \infty$. We consider $\mathcal{N}$ to be a sample of an underlying distribution when $\mathcal{N}$ is finite. In order to characterize this distribution we divide the line segment into pieces (cells) of length $\delta = 2^{-n}$, so that $N = 2^n$ cells are needed to cover $\mathcal{S}$. Here n is the *number of generations* in the binary subdivision of the line segment. We label the cells by the index $i = 0, 1, 2, \ldots, N-1$. The distribution of the population over the line is specified at the resolution δ, by the numbers, $\mathcal{N}_i$, of members of the population in the i-th cell. The fraction of the total population $\mu_i = \mathcal{N}_i/\mathcal{N}$ is a convenient *measure* for the content in cell i. The set $\mathcal{M}$, given by

$$\mathcal{M} = \{\mu_i\}_{i=0}^{N-1} \ , \tag{6.4}$$

gives a complete description of the distribution of the population at the resolution δ. The *measure* $M(\mathcal{L})$ of a part, or subregion $\mathcal{L}$, of the line segment $\mathcal{S}$ is

$$M(\mathcal{L}) = \sum_{i \in \mathcal{L}} \mu_i \ . \tag{6.5}$$

In general this is the end of the story; the only way one can describe the distribution of members over the line is by giving $\mathcal{M}$, with a suffi- cient resolution, that is, sufficiently small δ, to suit one's needs. However,

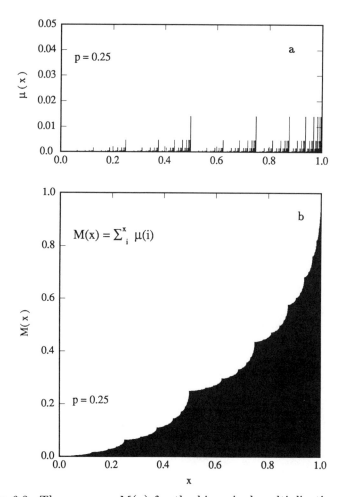

FIGURE 6.3: The measure $M(x)$ for the binominal multiplicative process after $n = 11$ generations with $\mu_0 = p = 0.25$ and $\mu_1 = 1 - p = 0.75$. (a) The measure or content μ of a cell as a function of the segment number (position) $x = i \cdot 2^{-11}$. (b) The measure $M(x)$ for the interval $[0, x]$ as a function of x.

when $\mathcal{M}$ has a scaling property, then much more can be said about the distribution, as we shall demonstrate using a simple example.

Consider the following *multiplicative process*, [1] which generates a measure on the unit interval $\mathcal{S} = [0, 1]$. First divide $\mathcal{S}$ into two parts of equal length $\delta = 2^{-1}$. The left half is given a fraction p of the population and

---

[1] This is the Besicovitch process discussed by Mandelbrot (1982, p. 377).

therefore the left segment has measure $\mu_0 = p$. The right-hand segment is given the remaining fraction and has the measure $\mu_1 = 1 - p$. Increase the resolution to $\delta = 2^{-2}$. The multiplicative process divides the population in each part in the same way. We find four pieces with the fractions of the population in the cells given by

$$\mathcal{M}_2 = \{\mu_i\}_{i=0}^{2^2-1} = \mu_0\mu_0 \,,\; \mu_0\mu_1 \,,\; \mu_1\mu_0 \,,\; \mu_1\mu_1 \,.$$

The next generation, $n = 3$, is obtained by dividing each cell into two new cells. A cell with the content μ_i is separated into a left-hand cell with measure $\mu_j = \mu_i\mu_0$ and a right-hand cell with measure $\mu_{j+1} = \mu_i\mu_1$. The whole line segment $[0, 1]$ is now divided into cells of length $\delta = 2^{-3}$, and the set $\mathcal{M}$ in the third generation is therefore given by the list of measures

$$\begin{aligned}\mathcal{M}_3 = \{\mu_i\}_{i=0}^{2^3-1} = \;&\mu_0\mu_0\mu_0 \,,\; \mu_0\mu_0\mu_1 \,,\; \mu_0\mu_1\mu_0 \,,\; \mu_0\mu_1\mu_1 \,,\\ &\mu_1\mu_0\mu_0 \,,\; \mu_1\mu_0\mu_1 \,,\; \mu_1\mu_1\mu_0 \,,\; \mu_1\mu_1\mu_1 \,.\end{aligned}$$

As this process is iterated, it produces shorter and shorter segments that contain less and less of the total measure. Figure 6.3 shows both the measure $\mu(x)$ of the cell located at x and the measure

$$M(x) = \sum_{i=0}^{x \cdot 2^n} \mu_i \,, \tag{6.6}$$

for the region $[0, x]$, after 11 generations of the multiplicative process. Here x specifies the cell index $i = x \cdot 2^n$. The measure $M(x)$ is scaling in the sense that the left half of figure 6.3 is obtained from the whole, and the right half from whole, by the relations (see also equation (10.10))

$$\begin{aligned}M(x) &= pM(2x) \,, &&\text{for } 0 \le x \le \tfrac{1}{2} \,,\\ M(x) &= p + (1-p)M(2x - 1) \,, &&\text{for } \tfrac{1}{2} \le x \le 1 \,.\end{aligned} \tag{6.7}$$

The relations (6.7) describe an *affine* transformation of the function $M(x)$, a notion which we discuss in some detail in chapter 10.

After n generations there are $N = 2^n$ cells labeled sequentially by the index $i = 0, \ldots, N - 1$. The length of the i-th cell is $\delta_n = 2^{-n}$, and the measure, or fraction of the population, in the cell is $\mu_i = \mu_0^k\mu_1^{(n-k)}$, where k is the number of 0's in the binary fraction representation of the number $x = i/2^n$. This can be seen if we represent the cell index i by a *binary fraction*:

$$x = i/2^n = \sum_{\nu=1}^{n} 2^{-\nu}\epsilon_\nu \,. \tag{6.8}$$

The 'digits,' ϵ_ν, have only two possible values: 0 and 1. For example, in the third generation, the first cell, $i = 0$, is represented by 0.000, the $i = 1$ cell by 0.001, the next cell by 0.010, etc., until finally the last cell, $i = 7$, is represented by 0.111. We need binary fractions having n digits in order to represent all the cells in the n-th generation.

Figure 6.3a shows the measure of the cells generated in the 11-th generation of the multiplicative process with $p = 0.25$. As shown in figure 6.3a there is one cell with the highest measure $(1 - p)^n$. There are 11 cells with measure $(1 - p)^{n-1}p^1$, etc. In general we have, with $\xi = k/n$ and $k = 0, 1, \ldots n$,

$$N_n(\xi) = \binom{n}{\xi n} = \frac{n!}{(\xi n)!((1 - \xi)n)!} \tag{6.9}$$

intervals with measure

$$\mu_\xi = \Delta^n(\xi), \quad \text{with} \quad \Delta(\xi) = \mu_0^\xi \mu_1^{(1-\xi)} = p^\xi(1 - p)^{(1-\xi)}. \tag{6.10}$$

The total measure of the segments representing the population is

$$M(x = 1) = \sum_{i=0}^{2^n - 1} \mu_i = \sum_{\xi=0}^{1} N_n(\xi)\Delta^n(\xi) = (\mu_0 + \mu_1)^n = 1. \tag{6.11}$$

The cells, describing the distribution of the population, cover the line completely, and contain all of the measure, that is, every member in the population.

6.3 Fractal Subsets

In the n-th generation $N_n(\xi)$ line segments have the length $\delta_n = 2^{-n}$ and the same measure μ_ξ. These segments form a subset, $\mathcal{S}_n(\xi)$, of the unit interval $\mathcal{S} = [0, 1]$. The points in the set $\mathcal{S}_n(\xi)$ have the same number, $k = \xi n$, of 0's among the n first decimal places in the binary expansion of the x-coordinate of the points. Of course, different points are represented by different sequences of zeros and ones. One finds, in the limit $n \to \infty$, that ξ is the fraction of zeros in the infinite binary fraction representation of the points in the set $\mathcal{S}_\xi$. This set is a *fractal* set of points. To see this we cover the set with intervals of length δ and form the d-measure $M_d(\mathcal{S}_\xi)$, as in equation (2.3), and determine the fractal dimension $D(\xi)$ of this set by studying how the M_d behaves as $\delta \to 0$, and $n \to \infty$:

$$M_d(\mathcal{S}_\xi) = \sum_{\mathcal{S}_\xi} \delta^d = N_n(\xi)\delta^d \xrightarrow[\delta \to 0]{} \begin{cases} 0, & d > D(\xi), \\ \infty, & d < D(\xi). \end{cases} \tag{6.12}$$

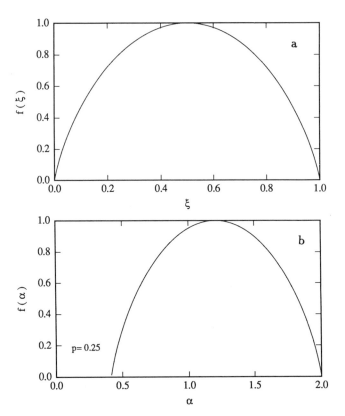

FIGURE 6.4: Fractal subsets of the measure generated by a binomial multiplicative process with $p = 0.25$. (a) The fractal dimension of subsets $\mathcal{S}_\xi$ of the interval that contain points x having a fraction ξ of zeros in the binary expansion of x, as a function of ξ. (b) The fractal dimension of subsets $\mathcal{S}_\alpha$ having a Lipschitz-Hölder exponent α, as a function of α.

We use Stirling's formula for $n!$:

$$n! = \sqrt{2\pi}\, n^{n+\frac{1}{2}}\, e^{-n} \,, \tag{6.13}$$

in equation (6.9) to find an approximate expression for $N_n(\xi)$:

$$N_n(\xi) \simeq \frac{1}{\sqrt{2\pi n \xi(1-\xi)}} \exp\{-n(\xi \ln \xi + (1-\xi)\ln(1-\xi))\} \,. \tag{6.14}$$

Noting that $n = -\ln \delta / \ln 2$, we find that the measure M_d in equation (6.12) may be written (neglecting the term $n^{-\frac{1}{2}}$, which gives only a logarithmic correction) as

$$M_d(\mathcal{S}_\xi) \sim \delta^{-f(\xi)} \delta^d \,, \tag{6.15}$$

with the exponent $f(\xi)$ defined by

$$f(\xi) = -\frac{\xi \ln \xi + (1 - \xi) \ln(1 - \xi)}{\ln 2} . \tag{6.16}$$

It follows that the d-measure M_d for the set $\mathcal{S}_\xi$ remains finite as $\delta \to 0$ only for $d = f(\xi)$, and therefore the fractal dimension, $D(\xi)$, of the set $\mathcal{S}_\xi$ is $f(\xi)$.

The population generated by the multiplicative process is spread over the set of points in the unit interval $\mathcal{S} = [0, 1]$. This set is a union of subsets $\mathcal{S}_\xi$:

$$\mathcal{S} = \bigcup_\xi \mathcal{S}_\xi . \tag{6.17}$$

The points in a subset $\mathcal{S}_\xi$ have, loosely speaking, the *same* population 'density.' The sets $\mathcal{S}_\xi$ are fractal with fractal dimensions $f(\xi)$ given by equation (6.16). The fractal dimension depends on the parameter ξ. We have plotted $f(\xi)$ as a function of ξ in figure 6.4a.

The measure $M(x)$ of the population distributed over the unit interval is completely characterized by the union of fractal sets. Each set in the union is fractal with its own fractal dimension. This is one reason for the term *multifractal*.

6.4 The Lipschitz-Hölder Exponent α

The parameter ξ is not very useful and in practice one uses the Lipschitz-Hölder exponent α instead (e.g., Mandelbrot, 1982, p. 373). The *singularities* of the measure $M(x)$ are characterized by α. Consider the measure generated by the multiplicative process at the n-th generation. This measure is a nondecreasing function of x, with increments $\mu_\xi = \Delta^n(\xi)$, at all x that have $\xi \cdot n$ zeros among the n first 'digits' when written as a binary fraction, that is, as $x = \sum_{\nu=1}^n 2^{-\nu} \epsilon_\nu$. Choose a $x(\xi)$ that corresponds to a given value of ξ; this point is a member of the set $\mathcal{S}_\xi$. The measure $M(x)$ is also given at a point $x(\xi) + \delta$, with $\delta = 2^{-n}$. The increment in $M(x)$ between these two points is μ_ξ and we have

$$\mu_\xi = M(x(\xi) + \delta) - M(x(\xi)) = \delta^\alpha , \tag{6.18}$$

where we have defined α by the equation

$$\mu_\xi = \delta^\alpha . \tag{6.19}$$

In the subsequent generations more and more points in the set $\mathcal{S}_\xi$ are obtained and equation (6.18) remains correct even in the limit $n \to \infty$.

A function $M(x)$ which satisfies equation (6.18) for all values of x has a derivative if $\alpha = 1$, is constant for $\alpha > 1$, and is *singular* if $0 \leq \alpha \leq 1$.

It follows from equations (6.10) and (6.19), and $\delta = 2^{-n}$, that the measure for a multiplicative population has a Lipschitz-Hölder exponent given by

$$\alpha(\xi) = \frac{\ln \mu_\xi}{\ln \delta} = -\frac{\xi \ln p + (1 - \xi) \ln(1 - p)}{\ln 2} . \tag{6.20}$$

This α holds for the points in the set $\mathcal{S}_\xi$, and is a linear function of ξ. α is a function also of the weight p, defining the subdivision of the interval. We find for the multiplicative measure with $p \leq \frac{1}{2}$ that $\alpha_{\min} \leq \alpha \leq \alpha_{\max}$ with

$$
\begin{aligned}
\alpha_{\min} &= -\ln(1 - p)/\ln 2 , &&\text{for } \xi = 0 , \\
\alpha_{\max} &= -\ln p/\ln 2 , &&\text{for } \xi = 1 .
\end{aligned}
\tag{6.21}
$$

There is a one-to-one correspondence between the parameters ξ and α, and therefore the subsets $\mathcal{S}_\xi$ may also be written as $\mathcal{S}_\alpha$. The measure $M(x)$ is characterized by the sets $\mathcal{S}_\alpha$, which as a union make up the unit interval $\mathcal{S} = [0, 1]$:

$$\mathcal{S} = \bigcup_\alpha \mathcal{S}_\alpha . \tag{6.22}$$

The measure has '*singularities*' with Lipschitz-Hölder exponent α on fractal sets $\mathcal{S}_\alpha$, which have the fractal dimension $f(\alpha) = f(\xi(\alpha))$. The $f(\alpha)$ curve for the measure of the population generated by the multiplicative process with $p = 0.25$, is shown in figure 6.4b.

In a recent paper Meneveau and Sreenivasan (1987) show that observations of fully developed turbulence are very well described by the process just discussed. The binomial multiplicative process, with $p = 0.7$, leads to an $f(\alpha)$ curve that describes accurately the observed multifractal spectrum of the dissipation field (see figure 6.5).

6.5   The $f(\alpha)$ Curve

The $f(\alpha)$ curve in figure 6.4b has a few quite general features which will be discussed further in section 6.8. The derivative of $f(\alpha)$ is

$$\frac{df(\alpha)}{d\alpha} = \frac{\ln \xi - \ln(1 - \xi)}{\ln p - \ln(1 - p)} . \tag{6.23}$$

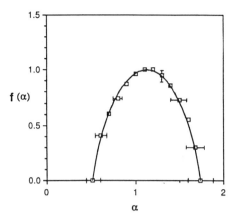

FIGURE 6.5: The multifractal spectrum for one-dimensional sections through the dissipation field in several fully developed turbulent flows (grid turbulence, wake of a circular cylinder, boundary layer, atmospheric turbulence). The symbols correspond to the experimental mean and the continuous curve is the $f(\alpha)$ curve for the binomial multiplicative process with $p = 0.7$ (Meneveau and Sreenivasan, 1987).

The maximum is $f(\alpha_0) = 1$, with

$$\xi = \tfrac{1}{2} \,,$$
$$f_{\max} = f(\alpha_0) = 1 \,, \qquad\qquad (6.24)$$
$$\alpha_0 = -\,\frac{\ln p + \ln(1 - p)}{2 \ln 2} \,.$$

It is a general result that the maximum value of the fractal dimension of the subsets $\mathcal{S}_\alpha$ equals the fractal dimension of the 'support' of the measure, which is 1 here since the measure is defined over the whole unit interval. For measures defined on fractals with a fractal dimension D one finds $f_{\max}(\alpha) = D$. Here the set $\mathcal{S}_{\alpha_0}$ has a fractal dimension of 1. This does not imply that this set covers the interval, but rather that $\mathcal{S}_{\alpha_0}$ contains a fraction of the points in the interval.

The maximum is at $\alpha_0 = 1.207\ldots$, for $p = 0.25$. The function $M(x)$ has zero derivative in the points where $\alpha > 1$. But $M(x)$ is a *singular* function because points where $\alpha(\xi) \leq 1$ are everywhere dense.

The discussion of the properties of the function $M(x)$ is somewhat delicate and involves questions such as whether or not the limit points of the sequence of points generated by the multiplicative process are included; see Mandelbrot (1982, p. 377), and Billingsley (1965). For a popular account of the properties of this function see Billingsley (1983). We discuss the function again in chapter 10. We comment here that the curve $M(x)$ has zero derivative 'almost everywhere.' Nevertheless it increases from 0 to 1, as x increases from 0 to 1. It is a *Devil's staircase*. The length of the curve from the origin to the end at (1,1) is equal to 2. The term 'almost everywhere' means here at all points except on a set of points with Lebesgue measure zero. These exceptional points can be covered by line segments of arbitrarily small total length. It is easy to see that the binary fractions have zero (linear) measure: Write the fractions in a systematic fashion $\frac{1}{2}, \frac{1}{4}, \frac{3}{4}, \frac{1}{8}, \frac{3}{8}, \ldots$. Cover the first point with a line segment of length δ^2, the next with a segment of length δ^3, the next with δ^4, and so on. The infinite sequence of these segments will cover all the binary fractions, and they have a length $\ell = \delta^2 + \delta^3 + \ldots = \delta^2/(1-\delta)$, which vanishes as $\delta \to 0$.

Another special point on the $f(\alpha)$ curve occurs at

$$
\begin{aligned}
\frac{f(\alpha)}{d\alpha} &= 1 \,, \\
\xi &= p \,, \\
f(\alpha_S) &= \alpha_S = S \,, \\
S &= -\{p \ln p + (1-p)\ln(1-p)\}/\ln 2 \,.
\end{aligned}
\tag{6.25}
$$

where a line through the origin is tangent to $f(\alpha_S)$. The fractal dimension of the set $\mathcal{S}_{\alpha_S}$ is S, which is recognized as the (information) entropy (e.g., Mandelbrot, 1982, p. 378) of the binomial multiplicative process. In the general multiplicative process, where the interval is subdivided into b cells with weights $p_0, p_1, \ldots, p_{b-1}$, one finds that the $f(\alpha_S)$ is given by

$$
S = -\sum_{\beta=0}^{b-1} p_\beta \log_b p_\beta \,,
\tag{6.26}
$$

where $\log_b$ is the base b logarithm. In the next section we show that almost all of the measure concentrates on the set $\mathcal{S}_{\alpha_S}$.

6.6   The Measure's Concentrate

The multiplicative process generates a population that has the overwhelming bulk of the initially uniform population concentrated into 'the set of concentration.' In the n-th generation the measure in the set $\mathcal{S}_n(\xi)$ is given by

$$
M(\mathcal{S}_n(\xi)) = N_n(\xi)\mu_\xi \simeq \frac{1}{\sqrt{2\pi np(1-p)}} \exp\left\{-\frac{n}{2p(1-p)}(\xi-p)^2\right\} \,.
\tag{6.27}
$$

The approximate expression on the right-hand side was obtained using equations (6.10) and (6.14) and expanding the exponential around the maximum value at $\xi = p$. The measure of the set $\mathcal{S}_n(\xi)$ as a function of ξ is very sharply peaked around $\xi = p$, and decreases as $n^{-\frac{1}{2}}$, with increasing n. However, a finite part, ϕ, of the total measure $M(x = 1)$ is contained in a union of sets

$$\mathcal{S}_\phi = \bigcup_{(p-\sigma) \leq \xi \leq (p+\sigma)} \mathcal{S}_n(\xi) \,, \tag{6.28}$$

with $\sigma \to 0$ as $n \to \infty$. To see this, note that the measure of the set $\mathcal{S}_\phi$ is given by (remember that $\xi = k/n$ so that $\sum_k \ldots \to n \int d\xi \ldots$)

$$M(\mathcal{S}_\phi) = \frac{n}{\sqrt{2n\pi p(1-p)}} 2 \int_p^{(p+\sigma)} d\xi \, \exp\left\{ -\frac{n(\xi - p)^2}{2p(1-p)} \right\}$$

$$= \frac{2}{\sqrt{\pi}} \int_0^\tau dt \, \exp\{-t^2\} \,, \tag{6.29}$$

with the upper limit of integration, τ, given by

$$\tau = \sigma \sqrt{\frac{n}{2p(1-p)}} \,. \tag{6.30}$$

This upper limit is determined by the condition $M(\mathcal{S}_\phi) = \phi$, i.e., that the measure of $\mathcal{S}_\phi$ is a finite fraction, $0 < \phi < 1$, of the total population. This gives $\tau = \tau(\phi)$, and we find

$$\sigma = \tau(\phi) \sqrt{\frac{2p(1-p)}{n}} \xrightarrow[n \to \infty]{} 0 \,. \tag{6.31}$$

This demonstrates that the measure *concentrates* on a set $\mathcal{S}_\phi$. In the limit $n \to \infty$ one finds that $\mathcal{S}_\phi$ is essentially $\mathcal{S}(\xi = p)$, with a fractal dimension given by the the entropy dimension S (Mandelbrot, 1982). This effect of concentration of the measure is called *curdling*.

Let us summarize. We find that a fraction of the *measure* arbitrarily close to 100% is contained in sets that have $\xi \simeq p$ for which $N(\xi)\mu_\xi$ is near its maximum. These sets have fractal dimensions given by the entropy of the multiplicative process. Of course, similar arguments lead to the conclusion that a finite fraction of the points in the interval is contained in sets with $\xi \simeq 1/2$ for which $N(\xi)$ is near its maximal value. These sets have fractal dimensions given by the fractal dimension of the support, which is 1 in this case.

6.7 The Sequence of Mass Exponents $\tau(q)$

Fractal structures observed experimentally, for example coastlines, or the viscous fingering pattern in figure 4.7, can also be modeled by numeric simulations as shown in figure 3.3. Both experimental observations and the results of simulations give sets of points $\mathcal{S}$, which then are presented in the form of curves or figures. Perhaps the most widely used method in the analysis of the structure of such sets is the *box-counting* method illustrated in figure 2.1. In this method the E-dimensional space of the observations is partitioned into (hyper-)cubes with side δ, and one counts the number $N(\delta)$ of cubes that contain *at least one* point of the set $\mathcal{S}$. Clearly, this is the crudest form of measure of the set, and it gives no information as to the *structure* of the set. For instance if the coastline folds back and forth so that it crosses a given 'box' a number of times n_i, that box still contributes only 1 to the number of boxes needed to cover the set — this somehow does not seem quite fair. Is there a way to give a higher weight to boxes with high n_i and lower weight to boxes with $n_i = 1$?

The answer has two main components: the *curdling* of the measure $M(x)$ on the set discussed in section 6.2 (Mandelbrot, 1974), and the unequal scaling ratios discussed in section 5.2. Grassberger (1983), Hentschel and Procaccia (1983b) and Grassberger and Procaccia (1983) used a measure that addresses the curdling problem. A related measure that also deals with the unequal scaling ratios has recently been introduced by Halsey et al. (1986b). These measures are identical, except for notation, to other probability measures discussed by Mandelbrot (1972, 1974, 1982) and by Voss (1985a). Meakin (1987b,c) has also discussed a related set of surface mass exponents.

A set $\mathcal{S}$ consisting of $\mathcal{N}$ points will have $\mathcal{N}_i$ points in the i-th cell. These points are sample points of an underlying measure. Let us use the 'mass' or *probability* $\mu_i = \mathcal{N}_i/\mathcal{N}$ in the i-th cell to construct the measure which may be written

$$M_d(q,\delta) = \sum_{i=1}^{N} \mu_i^q \delta^d = N(q,\delta)\delta^d \xrightarrow[\delta \to 0]{} \begin{cases} 0, & d > \tau(q) \ , \\ \infty, & d < \tau(q) \ . \end{cases} \qquad (6.32)$$

This measure has a *mass exponent* $d = \tau(q)$ for which the measure neither vanishes nor diverges as $\delta \to 0$. The mass exponent $\tau(q)$ for the set depends on the moment order q chosen. The measure is characterized by a whole sequence of exponents $\tau(q)$ that controls how the moments of the probabilities $\{\mu_i\}$ scale with δ. It follows from equation (6.32) that the

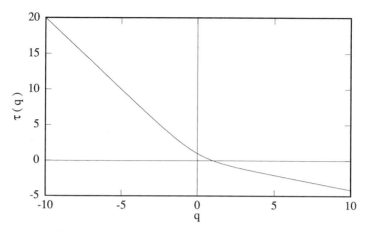

FIGURE 6.6: The sequence of mass exponents $\tau(q)$ as a function of moment order q, for the measure $M_d(q, \delta)$ for the binomial multiplicative process with $p = 0.25$.

weighted number of boxes $N(q, \delta)$ has the form

$$N(q, \delta) = \sum_i \mu_i^q \sim \delta^{-\tau(q)} , \tag{6.33}$$

and the mass exponent is given by

$$\tau(q) = - \lim_{\delta \to 0} \frac{\ln N(q, \delta)}{\ln \delta} . \tag{6.34}$$

We first note that if we choose $q = 0$ for the moment order q, then we have $\mu_i^{q=0} = 1$. Therefore we find that $N(q = 0, \delta) = N(\delta)$ is simply the number of boxes needed to cover the set, and $\tau(0) = D$ equals the fractal dimension of the set. The probabilities are normalized: $\sum_i \mu_i = 1$, and it follows from equation (6.34) that $\tau(1) = 0$.

Choosing large values of q, say 10 or 100, in equation (6.33) favors contributions from cells with relatively high values of μ_i since $\mu_i^q \gg \mu_j^q$, with $\mu_i > \mu_j$ if $q \gg 1$. Conversely, $q \ll -1$ favors the cells with relatively low values of the measure μ_i on the cell. These limits are best discussed by considering the derivative $d\tau(q)/dq$ given by

$$\frac{d\tau(q)}{dq} = - \lim_{\delta \to 0} \frac{\sum_i \mu_i^q \ln \mu_i}{(\sum_i \mu_i^q) \ln \delta} . \tag{6.35}$$

Let μ_- be the minimum value of μ_i in the sum. Then we find

$$\left. \frac{d\tau(q)}{dq} \right|_{q \to -\infty} = - \lim_{\delta \to 0} \frac{(\sum_i' \mu_-^q) \ln \mu_-}{(\sum_i' \mu_-^q) \ln \delta} ,$$

where the prime on the sum indicates that only cells with $\mu_i = \mu_-$ contribute. This expression may be rewritten as

$$\left. \frac{d\tau(q)}{dq} \right|_{q \to -\infty} = -\lim_{\delta \to 0} \frac{\ln \mu_-}{\ln \delta} = -\alpha_{\max} . \tag{6.36}$$

Here we have used the definition (6.19) of the Lipschitz-Hölder exponent α. A similar argument in the limit $q \to \infty$ leads to the conclusion that the minimum value of α is given by

$$\left. \frac{d\tau(q)}{dq} \right|_{q \to +\infty} = -\lim_{\delta \to 0} \frac{\ln \mu_+}{\ln \delta} = -\alpha_{\min} , \tag{6.37}$$

where μ_+ is the largest value of μ_i, which leads to the smallest value of α. In the next section we show that $\alpha = -d\tau/dq$ in general.

For $q = 1$ we find that $d\tau/dq$ has an interesting value:

$$\left. \frac{d\tau(q)}{dq} \right|_{q=1} = -\lim_{\delta \to 0} \frac{\sum_i \mu_i \ln \mu_i}{\ln \delta} = \lim_{\delta \to 0} \frac{S(\delta)}{\ln \delta} , \tag{6.38}$$

where $S(\delta)$ is the (information) *entropy* of the *partition* of the measure $\mathcal{M} = \{\mu_i\}_{i=0}^{N-1}$ over boxes of size δ, which may be written as

$$S(\delta) = -\sum_i \mu_i \ln \mu_i \sim -\alpha_1 \cdot \ln \delta . \tag{6.39}$$

The exponent $\alpha_1 = -(d\tau/dq)|_{q=1} = f_S$ is also the fractal dimension of the set onto which the measures concentrates and describes the scaling with the box size δ of the (partition) entropy of the measure. Note that the partition entropy $S(\delta)$ at resolution δ is given in terms of the entropy S of the measure by $S(\delta) = -S \ln \delta$ (see also equation (6.26)).

The general behavior of the sequence of mass exponents $\tau(q)$ is illustrated by the measure on the interval generated by the multiplicative binomial process. For this process we find that

$$N(d, \delta) = \sum_{k=0}^{n} \binom{n}{k} p^{qk} (1-p)^{q(n-k)} = (p^q + (1-p)^q)^n . \tag{6.40}$$

With the generation number given by $n = -\ln \delta / \ln 2$, as before, we find using equation (6.34) that

$$\tau(q) = \frac{\ln(p^q + (1-p)^q)}{\ln 2} \tag{6.41}$$

The resulting sequence of mass exponents is shown in figure 6.6. For $q = 0$, we find $\tau(0) = 1$, which is the dimension of the support, i.e., the unit interval.

6.8 The Relation between $\tau(q)$ and $f(\alpha)$

The sequence of mass exponents is related to the $f(\alpha)$ curve in a general way that is useful in applications. A multifractal measure is supported by a set $\mathcal{S}$, which is the union of fractal subsets $\mathcal{S}_\alpha$ with α chosen in the continuum of allowed values

$$\mathcal{S} = \bigcup_\alpha \mathcal{S}_\alpha . \tag{6.42}$$

What are the fractal dimensions describing the measure? Since the complete set $\mathcal{S}$ is fractal, with a fractal dimension D, the subsets have fractal dimensions $f(\alpha) \leq D$. For fractal subsets, with a fractal dimension $f(\alpha)$, the number $N(\alpha, \delta)$ of segments of length δ needed to cover the sets $\mathcal{S}_\alpha$ with α in the range α to $\alpha + d\alpha$ is

$$N(\alpha, \delta) = \rho(\alpha)d\alpha\, \delta^{-f(\alpha)} . \tag{6.43}$$

Here $\rho(\alpha)d\alpha$ is the number of sets from $\mathcal{S}_\alpha$ to $\mathcal{S}_{\alpha+d\alpha}$. For these sets the measure μ_α in a cell of size δ has the power-law dependence (6.2) on the length scale δ so that we may write $\mu_\alpha = \delta^\alpha$, and therefore the measure M for the set $\mathcal{S}$ given in equation (6.32) may be written

$$M_d(q, \delta) = \int \rho(\alpha)d\alpha\, \delta^{-f(\alpha)}\delta^{\alpha q}\delta^d = \int \rho(\alpha)d\alpha\, \delta^{q\alpha - f(\alpha)+d} . \tag{6.44}$$

The integral in equation (6.44) is dominated by the terms where the integrand has its maximum value, in other words for

$$\frac{d}{d\alpha}\{q\alpha - f(\alpha)\}\bigg|_{\alpha=\alpha(q)} = 0 . \tag{6.45}$$

The integral in equation (6.44) is therefore asymptotically given by

$$M_d(q, \delta) \sim \delta^{q\alpha(q)-f(\alpha(q))+d} . \tag{6.46}$$

Here M_d remains finite in the limit $\delta \to 0$ if d equals the mass exponent $\tau(q)$ given by

$$\tau(q) = f(\alpha(q)) - q\alpha(q) , \tag{6.47}$$

where $\alpha(q)$ is the solution of equation (6.45). Thus the mass exponent is given in terms of the Lipschitz-Hölder exponent $\alpha(q)$ for the mass, and the fractal dimension $f(\alpha(q))$ of the set that supports this exponent.

We may, on the other hand, if we know the mass exponents $\tau(q)$, determine the Lipschitz-Hölder exponent and f using equations (6.45) and

q	$\tau(q)$	$\alpha = -d\tau(q)/dq$	$f = q\alpha + \tau(q)$
$q \to -\infty$	$\sim -q\alpha_{max}$	$\to \alpha_{max} = -\ln\mu_-/\ln\delta$	$\to 0$
$q = 0$	D	α_0	$f_{max} = D$
$q = 1$	0	$\alpha_1 = -S(\delta)/\ln\delta$	$f_S = \alpha_1 = S$
$q \to +\infty$	$\sim -q\alpha_{min}$	$\to \alpha_{min} = -\ln\mu_+/\ln\delta$	$\to 0$

TABLE 6.1: Special values and limits of the sequence of mass exponents $\tau(q)$ and of the $f(\alpha)$ curve for a multifractal measure, $\mathcal{M} = \{\mu_i\}$, supported by a set with fractal dimension D. Here q is the moment order of $\mathcal{M}$ (see equation (6.32)). The largest and smallest probabilities in boxes of size δ are μ_- and μ_+, respectively. $S(\delta)$ is the entropy of the partition of the measure $\mathcal{M}$ over boxes of size δ. The measure has the entropy $S = -\lim_{\delta \to 0} S(\delta)/\ln\delta = f_S$, which is the fractal dimension of the set of concentration for the measure.

(6.47). This gives

$$\alpha(q) = -\frac{d}{dq}\tau(q) \,,$$

$$f(\alpha(q)) = q\alpha(q) + \tau(q) \,. \tag{6.48}$$

These two equations give a parametric representation of the $f(\alpha)$ curve, i.e., the fractal dimension, $f(\alpha)$, of the support of 'singularities' in the measure with Lipschitz-Hölder exponent α. The $f(\alpha)$ curve characterizes the measure and is equivalent to the sequence of mass exponents $\tau(q)$. The pair of equations (6.48) in effect constitute a Legendre transformation from the independent variables τ and q to the independent variables f and α. Using the pair of equations 6.48 for the simple example of the binomial multiplicative process with $\tau(q)$ given by equation (6.41) (see figure 6.6), we recover the $f(\alpha)$ curve shown in figure 6.4b.

The maximum of the $f(\alpha)$ curve occurs for $df(\alpha)/d\alpha = 0$. From equation (6.45) it follows that we then have $q = 0$, and we conclude from equation (6.48) that $f_{max} = D$, since we have shown that $\tau(0) = D$, where D is the fractal dimension of the support of the measure. We have summarized the various relations between the $f(\alpha)$ curve and the sequence of mass exponents in the table above.

6.9   Curdling with Several Length Scales

Consider the generalization of the binomial multiplicative measure to a Cantor set support as illustrated in figure 6.7. In each generation each surviving piece is divided into two intervals, a small piece of relative length $l_0 = 0.25$ and a large piece of relative length $l_1 = 0.4$, whereas the middle piece is cut out. The small piece is given a fraction $p_0 = 0.6$ and the large piece is given only a fraction $p_1 = 0.4$ of the measure contained in the parent interval.

If we try to apply the definition equation (6.32) to the Cantor bar shown in figure 6.7, we immediately see that the definition is inadequate since it uses the *same* box size δ on all cells used to cover the set. The measure recently used by Halsey et al. (1986b) is a combination of the two generalizations (5.4) and (6.32). Consider, as in the definition (5.4), that the fractal set $\mathcal{S}$ may be partitioned into some number N of disjoint pieces $\mathcal{S}_0, \mathcal{S}_1, \ldots, \mathcal{S}_{N-1}$, covering the set. Let the Euclidean length of the i-th set be l_i so that $\mathcal{S}_i$ fits into a (hyper-)cube of side l_i such that $l_i < \delta$, for all i. The measure is then defined by

$$M_d(q, \delta) = \sum_{i=0}^{N-1} \mu_i^q l_i^d \xrightarrow[\delta \to 0]{} \begin{cases} 0, & d > \tau(q) \,, \\ \infty, & d < \tau(q) \,. \end{cases} \qquad (6.49)$$

This measure again has a mass exponent $d = \tau(q)$ for which the measure neither vanishes nor diverges as $\delta \to 0$.

We illustrate the use of this measure by a discussion of the two-scale ($l_0 = 0.25$, $l_1 = 0.4$) Cantor set, with a measure generated by a multiplicative process ($p_0 = 0.6$, $p_1 = 0.4$) (see figure 6.7, and Halsey et al., 1986b). Since we have $\binom{n}{k}$ segments of length $l_0^k l_1^{n-k}$ and weight $\mu_i = p_0^k p_1^{n-k}$ in the n-th generation of the Cantor bar in figure 6.7, we find that the measure is easily evaluated and given by

$$M_d(q, \delta) = \sum_{k=0}^{n} \binom{n}{k} (p_0^q l_0^d)^k (p_1^q l_1^d)^{n-k} = (p_0^q l_0^d + p_1^q l_1^d)^n \,. \qquad (6.50)$$

This measure remains finite with $\delta = l_1^n$ as $n \to \infty$ if, and only if, one chooses $d = \tau(q)$, where $\tau(q)$ is the solution of the equation

$$p_0^q l_0^{\tau(q)} + p_1^q l_1^{\tau(q)} = 1 \,. \qquad (6.51)$$

We have solved the equation (6.51) for $\tau(q)$ numerically. Once $\tau(q)$ is available the $f(\alpha)$ curve shown in figure 6.8 is obtained as before (see equation (6.48)).

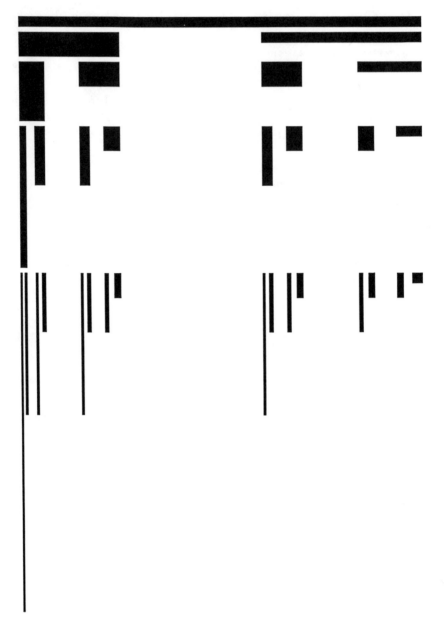

FIGURE 6.7: A two-scale fractal measure on a Cantor bar with $l_0 = 0.25$ and weight $p_0 = 0.6$, and $l_1 = 0.4$ with $p_1 = 0.4$. The height of the bars in the n-th generation is proportional to the density $\rho_i = \mu_i/l_i$. The fractal dimension of the Cantor dust supporting the measure is $D = 0.6110$.

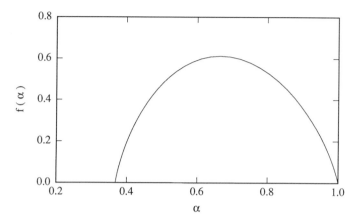

FIGURE 6.8: The $f(\alpha)$ curve for the two-scale fractal measure on a Cantor bar $\mathcal{S}$, with $l_0 = 1/4$, $l_1 = 2/5$ and weights $p_0 = 3/5$ and $p_1 = 2/5$. This represents the fractal dimension, f, of the subsets $\mathcal{S}_\alpha$ with Lipschitz-Hölder exponent α as function of α.

For $q = 0$ the mass exponent $\tau(0)$ equals $D = 0.6110$ — the same value we determined for the fractal dimension of the set using the definition (5.4) for the measure of the set. In the limit $q \to \infty$, we have that $p_0^q \gg p_1^q$. Therefore the first term dominates the left-hand side in equation (6.51) and $\tau(q)$ is simply given by the equation

$$p_0^q l_0^{\tau(q)} = 1 .$$

This leads to the result

$$\tau(q)|_{q \to +\infty} = -q\alpha_{min} , \qquad \text{with} \quad \alpha_{min} = \frac{\ln p_0}{\ln l_0} = 0.3685 . \qquad (6.52)$$

Similarly, as $q \to -\infty$, the p_1 term dominates and we find that

$$\tau(q)|_{q \to -\infty} = -q\alpha_{max} , \qquad \text{with} \quad \alpha_{max} = \frac{\ln p_1}{\ln l_1} = 1 . \qquad (6.53)$$

Halsey et al. (1986b) discussed the measure given in equation (6.49) in terms of the dimension function D_q introduced by Grassberger (1983), Hentschel and Procaccia (1983b) and Grassberger and Procaccia (1983), which is given by

$$D_q = \tau(q)/(1 - q) , \qquad (6.54)$$

where the numeric factor $(1 - q)$ modifies the mass exponent $\tau(q)$ to give the result that $D_q = E$ for sets of constant density in E-dimensional space.

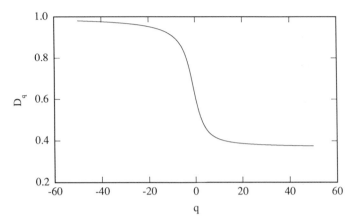

FIGURE 6.9: The spectrum of fractal dimensions D_q as a function of the moment order q for the triadic Cantor dust with $l_0 = 1/4$, $l_1 = 2/5$ and weights $p_0 = 3/5$ and $p_1 = 2/5$. $D_\infty = \ln p_0 / \ln l_0 = 0.3685$ and $D_{-\infty} = \ln p_1 / \ln l_1 = 1$.

The spectrum of fractal dimensions D_q i then given by

$$D_q = \frac{1}{q-1} \lim_{\delta \to 0} \frac{\ln N(q,\delta)}{\ln \delta} . \tag{6.55}$$

The moment order q is any number in the range $-\infty$ to ∞, and the function D_q is the *spectrum of fractal dimensions* for a *fractal measure* on the set S. If we choose $q = 0$ for the moment order q, then we have $\mu_i^{q=0} = 1$. Therefore we find that $N(q = 0, \delta) = N(\delta)$ is simply the number of boxes needed to cover the set, and $D_{q=0} = D$ equals the fractal dimension of the set. In the limit $q \to \infty$, we find that $D_\infty = \alpha_{\min}$, whereas $D_{-\infty} = \alpha_{\max}$.

Because of the singularity, $1/(1-q)$, in the definition of D_q, we must take a little care in the evaluation of $N(q, \delta)$ for $q \to 1$. Noting that $\mu_i^q = \mu_i \mu_i^{q-1} = \mu_i \exp\{(q-1)\ln \mu_i\}$, and using that $\exp\{(q-1)\mu_i\} \to 1 + (q-1)\ln \mu_i$, in the limit $q \to 1$, we find that

$$\ln \left(\sum_i \mu_i^q \right) \to \ln \left\{ 1 + (q-1) \sum_i \mu_i \ln \mu_i \right\} \simeq (q-1) \sum_i \mu_i \ln \mu_i$$

Here we have used $\sum_i \mu_i = 1$. We therefore find that $D_{q=1}$ is given by

$$D_1 = \lim_{\delta \to 0} \frac{\sum_i \mu_i \ln \mu_i}{\ln \delta} . \tag{6.56}$$

This dimension describes the scaling behavior of the partition *entropy* of the measure on the set S. The *entropy* $S(\delta)$ defined in statistical physics for probabilities $\{\mu_i\}$ distributed over boxes of size δ is given by

$$S = -\sum_i \mu_i \ln \mu_i \sim -D_1 \ln \delta . \tag{6.57}$$

In terms of the relations derived in the previous section we conclude that for $q = 1$ we find $\alpha = f = D_1$.

In order to get some feeling for the spectrum of dimensions let us consider the normal Euclidian case. For a uniform measure in E-dimensional space with a constant density of points, we divide space into $N = \delta^{-E}$ cells with a volume of δ^E each. Then $\mu_i = \delta^E$, and we find

$$\sum_{i=1}^{N} \mu_i^q = \sum_{i=1}^{N} \delta^{qE} = \delta^{(q-1)E} .$$

We find, using this result in the definition (6.55), that the spectrum of fractal dimension D_q for a uniform measure in space equals the Euclidian dimension:

$$D_q = \frac{1}{q-1} \lim_{\delta \to 0} \frac{\ln \delta^{(q-1)E}}{\ln \delta} = E . \tag{6.58}$$

Therefore we have the result that the spectrum of fractal dimensions is E for a uniform measure and is independent of the moment order q, i.e., it has no structure. This result also explains the use of the factor $(1-q)$ in the definition of D_q. In passing, we note that the entropy is $S = -\sum_i \mu_i \ln \mu_i = \ln \mathcal{N} = -E \ln \delta$, so that the entropy dimension is indeed E.

In figure 6.9 we have plotted the spectrum of fractal dimensions D_q as a function of the moment order q for the Cantor bar shown in figure 6.7.

6.10   Multifractal Rayleigh-Benard Convection

A remarkable application of these ideas to observations of the hydrodynamic instability in thermal convection has been made by Jensen et al. (1985). They studied the thermal convection of mercury in a small cell 0.7 cm × 1.4 cm in area and 0.7 cm high. With the bottom temperature fixed at $T_{\text{bottom}} = T_{\text{top}} + \Delta T$, that is, above the temperature at the top surface, convection in the form of two rolls with a horizontal axis occurs.

Increasing the temperature difference ΔT beyond some critical value results in an instability of the convection rolls, which start to bend with a frequency $\omega_0 \simeq 230$ mHz. Jensen et al. perturbed this system by placing the cell in a horizontal magnetic field and sending an electric ac current

FIGURE 6.10: The experimental attractor for thermal convection. The time series of temperatures, x_t, is used to plot the temperature at time $t+1$, i.e. x_{t+1}, versus x_t. The figure corresponds to a projection of the attractor of the motion in phase space (Jensen et al., 1985).

between the bottom and top at the center of the cell. The frequency of the ac current is $\omega_{ac} = \Omega\omega_0$. The ratio of the two frequencies Ω is the winding number and is chosen to be an irrational number — the golden mean: $\Omega = \Omega_{gm} = (\sqrt{5} - 1)/2$. The reason for this particular choice is found in dynamic systems theory.

The result of this perturbation is that the temperature measured with a thermometer near the bottom of the cell fluctuates irregularly in time. A record of these fluctuations is a time series of temperatures, x_t, where t is time in units of the time interval between observations. In figure 6.10 a plot of x_{t+1} as a function of x_t is shown for 2500 observations of x_t. The set of points reflects the strange attractor in phase space for the motion of this system. The points on the attractor representing the time series are concentrated with various intensities in different regions. Some of the bunching of points observed is due to the fact that the points in figure 6.10 represent a projection of the true attractor in phase space. However, by using a three-dimensional space with coordinates (x_t, x_{t+1}, x_{t+2}) to represent the time series, this projection effect becomes negligible, and the experimental data are in fact analyzed in this space.

Jensen et al. analyzed the experimental results as follows. Start from a given point x_t on the orbit in figure 6.10 and count the number m_t of steps along the time series required before the point returns to within a distance δ of the starting point. An estimate of the probability p_t of being in a hyper-cube of side δ is $p_t \sim 1/m_t$. Now evaluate the measure

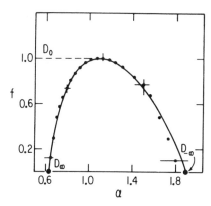

FIGURE 6.11: The $f(\alpha)$ curve for thermal convection. The points are estimated from the observed time series. The curve is the $f(\alpha)$ curve for the critical circle map. Note that there are *no* free parameters and the observations fit the theory remarkably well (Jensen et al., 1985).

$M_d(q, \delta)$ given in equation (6.49). Remember, however, that the sum in equation (6.49) is over the boxes needed to cover the set of points — not a sum over the points. This is easily corrected because the number of points in the i-th cell is $\sim \sum_{t \in i} 1/p_t$. From the experimental points one obtains

$$M_d(q, \delta) \sim \delta^d \sum_t \left(m_t(\delta)\right)^{1-q} . \qquad (6.59)$$

The measure in equation (6.59) is evaluated for cell sizes δ that vary over two orders of magnitude and the critical dimension is obtained from a log–log plot of the relation

$$\sum_t m_t(\delta)^{(1-q)} \sim \delta^{-\tau(q)} . \qquad (6.60)$$

Having obtained the critical dimension one finds $\alpha(q)$ and $f(q)$ from the equations (6.48). The experimental points obtained in this way are plotted in figure 6.11. The error bars on the experimental points have been obtained by varying the range in δ over which the power law in equation (6.60) is fitted to the observations. The general shape of the observed $f(\alpha)$ curve is similar to the curve for the two-scale Cantor bar shown in figure 6.8. Note that this analysis uses more of the experimental information than is contained in figure 6.10 since p_t explicitly depends on the time-sequence of points in figure 6.10.

The truly remarkable fact is that the curve in figure 6.11 is *not* a fit to the observations but an independently calculated curve for the *circle map* with a golden mean winding number!

The circle map is an iterative mapping of one point on a circle to another. Specify points on a circle by giving the angle θ to that point measured from some direction. Start with an arbitrary point θ_0 and generate a series of points by the repeated application of the mapping

$$\theta_{n+1} = \theta_n + \Omega - \frac{K}{2\pi} \sin 2\pi\theta_n \ . \tag{6.61}$$

This mapping has been studied in great detail (see for example Jensen et al., 1983, 1984a,b) and it has a strange attractor for the critical cycle obtained with the critical value $K = 1$ and with a winding number $\Omega = \Omega_{gm}$. The curve in figure 6.11 is the $f(\alpha)$ curve for this circle map.

The results in figure 6.11 show that thermal convection perturbed at the golden mean winding number and the critical circle map have the same fractal structure, and therefore belong to the same *universality class*. In addition, the fractal dimension of the set is the maximum value found for f, giving $f = D_0 = 1$, which is expected since the support of the measure is the circle, which is one-dimensional. For a further discussion see also Procaccia (1986).

In conclusion we remark that the determination of the proper p_i's is by no means a trivial enterprise. In fact, the determination of the correct procedure for the determination of the measure equation (6.32) for an experimental set is analogous to identifying the *order parameter* for a phase transition. Once the order parameter for a phase transition is identified the whole machinery of the Landau theory of phase transitions applies, and the critical behavior may be calculated using renormalization group methods. However, in order to determine the order parameter a deep insight into the phenomena at hand is required.

6.11   DLA and the Harmonic Measure

Consider the cluster generated by the diffusion limited-aggregation (DLA) process in figure 3.3. How can we best characterize the surface or perimeter of such fractal structures? The *harmonic measure* affords a method of quantitatively characterizing such surfaces. This (probability) measure is defined (with respect to a particular cluster) as the probability, $p(\mathbf{r})d\mathbf{r}$, of a random walker approaching the cluster from infinity first striking the cluster between the points $\mathbf{r}$ and $\mathbf{r} + d\mathbf{r}$ along the boundary of the cluster.

In practice one estimates the harmonic measure using simulations. First a DLA cluster is grown by the Witten-Sander algorithm and the growth is *stopped* when the cluster contains a large number N of particles and has a diameter L. The DLA cluster shown in figure 6.12a is typical.

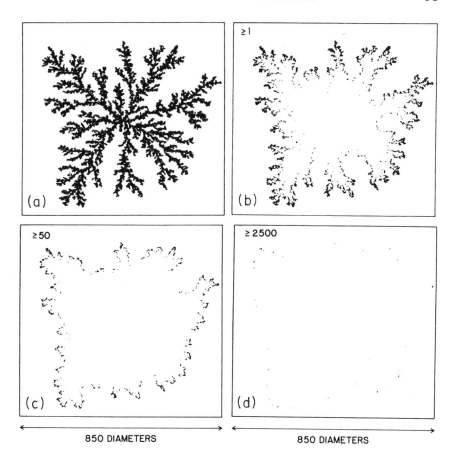

850 DIAMETERS   850 DIAMETERS

FIGURE 6.12: The result of a simulation in which 10^6 random walk particles were used to probe the surface of a two-dimensional off-lattice DLA cluster. After a particle had contacted the cluster, it was removed and a new particle started on a random walk trajectory from a random position on a circle enclosing the cluster. (a) The DLA cluster containing 50,000 particles. (b–d) show the location of the particles in the cluster which were contacted at least once, ≥ 50 times and ≥ 2500 times (Meakin et al., 1986).

It has no loops, and the number of sites in the perimeter, N_P, i.e., the number of possible growth sites, is proportional to the number, N, of particles in the cluster. They both scale with the radius of gyration R_g, or the diameter L of the cluster:

$$N_P \sim N \sim L^D . \tag{6.62}$$

Here the fractal dimension of DLA clusters is $D = 1.71$ in two dimensions.

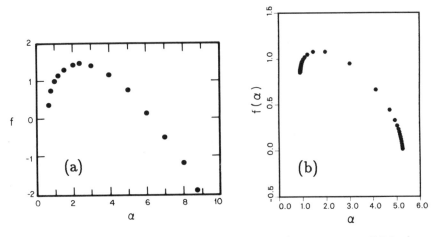

FIGURE 6.13: (a) The $f(\alpha)$ curve for the harmonic measure on DLA clusters. The largest cluster used in the estimate contained 150 particles (Amitrano et al., 1986). (b) The $f(\alpha)$ curve for the harmonic measure *calculated* for a viscous fingering structure (similar to the one shown in figure 4.5b) observed in a Hele-Shaw cell (Nittmann et al., 1987).

The possible growth sites on the perimeter of the DLA cluster are numbered with the index $k = 1, \ldots, N_P$.

The harmonic measure is estimated for the given DLA cluster in a further simulation in which a large number $\mathcal{N}$ of random walk particles are used to probe the surface of the DLA cluster. After a particle has contacted the cluster, it is removed and a new particle is started on a random walk trajectory from a random position on a circle enclosing the cluster. The probability p_k of a random walker contacting the k-th site on the perimeter is estimated to be $p_k = \mathcal{N}_k/\mathcal{N}$, where $\mathcal{N}_k$ is the number of times the k-th site was contacted. The set of probabilities

$$\mathcal{H} = \{p_k\}_{k=1}^{N_P} \tag{6.63}$$

represents the increments of the harmonic measure $M_\mathcal{H}$ at the resolution δ corresponding to the diameter of the diffusing particles. For recent discussions of this measure see Meakin (1987b,c) and Hayakawa et al. (1987).

The spectrum of fractal dimensions for the harmonic measure is obtained as in section 6.7. However, a slight modification is required because one uses a fixed particle diameter δ, and studies the effect of increasing the diameter L of the cluster. Equation (6.33) then becomes

$$N(q, L) = \sum_k p_k^q \sim \left(\frac{\delta}{L}\right)^{-\tau(q)} \sim L^{\tau(q)} . \tag{6.64}$$

Makarov (1985) has shown that the information dimension, $f(q = 1)$, is exactly 1 for the harmonic measure on the boundary of *any* connected domain in two dimensions. Simulations are required for other values of q. The tips have the highest probabilities and dominate the sum in equation (6.64), for $q \gg 1$. Accurate estimates of D_q may be obtained in this range. However, for $q \ll -1$, one finds that the sum is dominated by the smallest probabilities and they are very difficult to estimate accurately since the probability of a random walker reaching the bottom of the 'fjords' is practically zero. Once $\tau(q)$ is obtained from the simulations one finds the Lipschitz-Hölder exponent α, and the fractal dimension $f(\alpha)$ of the subsets S_α where the harmonic measure has singularities with exponent α, using the pair of equations (6.48). Amitrano et al. (1986) used an electrostatic formulation of the DLA problem and solved the resulting equations numerically. The $f(\alpha)$ curve they obtained is shown in figure 6.13a. Note that $f(\alpha)$ is the fractal dimension of a subset of the sites that support the harmonic measure and therefore we expect $f(\alpha) \geq 0$. The negative values of $f(\alpha)$ in figure 6.13a are an artifact that results from averaging many independent simulations for the probabilities p_k (see also Meakin, 1987b,c).

The harmonic measure gives a useful characterization of the complicated surface of fractal clusters. One expects the maximum of the $f(\alpha)$ curve to occur at $q = 0$. But $p_k^0 = 1$ and one finds that $N(q = 0, L)$ is simply the number of perimeter sites N_P, and therefore $f_{\max} = D = 1.71$. The simulations by Amitrano et al. (see figure 6.13a) gave $f_{\max} \simeq 1.5$. This low value may be attributable to the fact that they could handle only rather small clusters containing ≤ 150 particles. The maximum value of f, estimated from the analysis of the harmonic measure for the viscous fingering structure (figure 6.13b), is even lower.

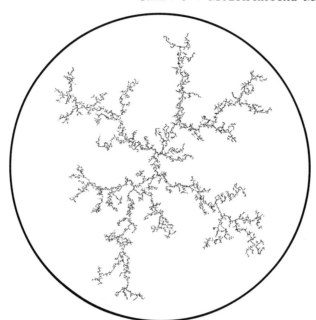

FIGURE 6.14: Fractal viscous fingering with a fractal dimension $D = 1.64 \pm 0.04$ in a two-dimensional porous medium consisting of a layer of 1-mm glass spheres placed at random and sandwiched between two plastic sheets. Air (black) displaced glycerol at a capillary number Ca of 0.15. The structure was observed at a time $t = 0.8\,t_0$, where the breakthrough time is $t_0 = 28.6$ s (Måløy et al., 1987b).

6.12   Multifractal Growth of Viscous Fingers

The viscous fingers that occur in the displacement of high-viscosity fluids by low-viscosity fluids in porous media generate fractal structures (see figure 4.7) that bear a remarkable resemblance to the fractals that arise in DLA (see figure 3.3). Recently it was shown that the *dynamics* of DLA and viscous fingering is also the same. The length of the longest finger and the radius of gyration as a function of time are the same for the DLA simulations and the experimental observations on two-dimensional viscous fingering in porous media (Meakin, 1987a; Måløy et al., 1987a,b).

The multifractal nature of the harmonic measure for DLA clusters was discussed in the previous section. The distribution of pressure gradients at the surface of the growing viscous fingering structure corresponds to the harmonic measure in DLA. However, this pressure gradient is not accessible experimentally. We have introduced a *new* measure, the *new-growth measure*, that gives a multifractal characterization of the growth dynamics of viscous fingers (Måløy et al., 1987b).

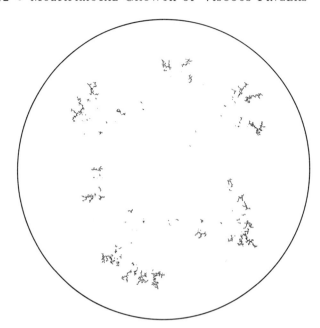

FIGURE 6.15: Active growth zone of the viscous fingering structure shown in figure 6.14. The time between the two pictures used to construct the growth zone was 2.8 s, which corresponds to a relative time increment of $\Delta t/t_0 = 0.10$ (Måløy et al., 1987b).

Figure 6.14 shows a viscous fingering structure obtained in an experiment in which air displaces glycerol at a high capillary number. The viscous fingering structure grows mainly at the tips. The new pores invaded in a small time interval as shown in figure 6.15.

In our experiments we measure the 'mass' m_i of the growth 'islands' shown in figure 6.15. The islands are numbered in an arbitrary way by the index $i = 1, 2, \ldots, N_I$, where N_I is the number of sites at which we observe growth. Let the total mass of the islands be $m_0 = \sum_i^{N_I} m_i$, and introduce the normalized mass μ:

$$\mu_i = \frac{m_i}{\sum m_i} = \frac{m_i}{m_0} . \tag{6.65}$$

The set $\mathcal{M} = \{\mu_i\}$ characterizes the *observed* growth of the structure. This set represents the increments in the underlying *new-growth measure* $M_{\mathcal{M}}$ at the resolution of the experiment. This new measure is to be distinguished from the *harmonic measure* $M_{\mathcal{H}}$ discussed in section 6.11, in the context of DLA.

Any experimentally observed viscous finger structure is a realization of a stochastic process. At any instant the growing finger structure may invade any pore on the perimeter of the structure. We label the perimeter sites with the index k. The dynamics of the finger growth is then in principle controlled by the set $\mathcal{H} = \{p_k\}$ of probabilities that the k-th pore on the perimeter is invaded next. The set $\mathcal{H}$ of growth probabilities are the increments in the harmonic measure $M_\mathcal{H}$ of the viscous fingering structure, at the resolution of the experiment. The measure $M_\mathcal{H}$ *changes* as soon as a pore is invaded since the perimeter of the structure changes. The actual growth at any site changes the growth probabilities at all the other sites. The *new-growth measure* $M_\mathcal{M}$ expresses the integrated effect of the sequence of pore invasion processes and we conclude that $M_\mathcal{M}$ is only indirectly related to the harmonic measure.

In the following sections we analyze experiments on viscous fingering in terms of the fractal new-growth measure.

The Fractal Set of Growth Sites

Consider the set of points $\mathcal{N}$ at which we have observed growth, i.e., the set of pores from which further growth has occurred. The points in this set have $\mu_i > 0$, and the set $\mathcal{N}$ is the old-growth – new-growth interface. The number of points in $\mathcal{N}$ is N_I and increases with the size of the viscous fingering structure. For a fractal structure we expect N_I to be given by

$$N_I = a \left(\frac{R_g}{\delta} \right)^{D_I} . \qquad (6.66)$$

Here D_I is the dimension of the growing interface, δ is the pixel size at which the structure is analyzed and R_g is the radius of gyration. We could equally well have used the length of the longest finger instead of the gyration radius R_g, because these lengths are proportional within the experimental uncertainties. Equation (6.66) gives an increase in N_I with increasing size of the growing finger structure, and a decrease in N_I, as the pixel size δ is increased, decreasing the resolution at which the set of points $\mathcal{N}$ is analyzed.

We have counted N_I for three sequences of pictures of our experiments. In figure 6.16 we have plotted N_I as a function of the corresponding radius of gyration R_g on a log-log plot. From the fits in figure 6.16 we find that the set of points $\mathcal{N}$ is fractal with a dimension

$$D_I = 1.0 \pm 0.1 , \qquad (6.67)$$

and with an amplitude $a = 1.1 \pm 0.5$. The old-growth – new-growth interface is a *fractal dust* in the plane, and has a fractal dimension of 1.

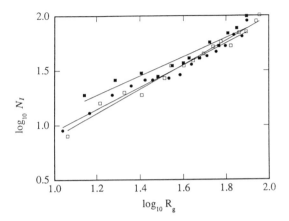

FIGURE 6.16: The number growth sites as a function of the radius of gyration for three experiments in which air displaces glycerol. The straight lines represent fits of equation (6.66) to the observed values. The parameters of the fit are : • $- a = 0.9 \pm 0.2$, $D_I = 1.01 \pm 0.05$. ■ $- a = 1.7 \pm 0.6$, $D_I = 0.87 \pm 0.08$. □ $- a = 0.6 \pm 0.1$, $D_I = 1.11 \pm 0.04$. The fractal dimension of the growing interface is $D_I = 1.0 \pm 0.1$.

The box-counting dimension of the set of points $\mathcal{N}$ at a given radius of gyration is obtained by changing the box size δ. We find by fitting equation (6.66) to the observed box counts, for intermediate box sizes, an estimate for the old-growth – new-growth interface fractal dimension: $D_I = 1.0 \pm 0.2$. The estimate is uncertain because the observed set of points $\mathcal{N}$ is only a finite *sample* of the underlying fractal measure. Of course, for any finite set of points one finds that $N_I(\delta \to 0) = $ constant, corresponding to dimension zero. The crossover to $D = 0$ is simply due to the finite resolution of our observations.

Meakin and Witten (1983) and Meakin et al. (1985) have studied the growing interface in DLA simulations and found that N_I increases as $N_0^{0.625 \pm 0.02}$, where N_0 is the cluster mass. This result implies a dimension of $D_I = 1.07 \pm 0.04$, since $N_0 \sim R_g^D$, where $D = 1.71$ is the fractal dimension of DLA clusters. Meakin and Witten (1983) also determined the number of particles N_I touching an initial cluster of size N_0 after K particles had been added to the cluster. They found that the number N_I approached a limit when $K \to \infty$, so that N_I depends on the cluster structure alone and does not depend on K in the limit of large K.

We have a satisfying agreement between experiment and simulation and we conclude that both experiments and simulations give $D_I \simeq 1$ as the fractal dimension of the old-growth – new-growth interface.

The $f(\alpha)$ Curve

So far we have only discussed the set of points at which $\mu_i > 0$. We may specify a *subset* $\mathcal{N}_\mu$, consisting of all the growth sites for which $\mu \leq \mu_i \leq \mu + \Delta\mu$. If we specify the measure in a scale-independent way we may find that such subsets are fractal sets. We therefore choose to specify subsets of the growth sites by a Lipschitz-Hölder exponent α, as in equation (6.19):

$$\mu = \left(\frac{\delta}{R_g}\right)^\alpha . \tag{6.68}$$

This relation simply gives a definition of α:

$$\alpha = \frac{\ln\mu}{\ln(\delta/R_g)} . \tag{6.69}$$

We could, of course, equally well choose the length of the longest finger instead of R_g in the definition of α. This would only correspond to a shift in α that is irrelevant for large clusters.

Let us choose α to be in the range α to $\alpha + \Delta\alpha$. From equation (6.68) we then find the corresponding range of μ_i for a finger structure having a radius of gyration R_g, observed at a resolution δ. The set of growth sites that have islands that give μ_i in the specified range form a set of points $\mathcal{N}_\alpha$. The set of all growth sites may then be written as the union of such sets:

$$\mathcal{N} = \bigcup_\alpha \mathcal{N}_\alpha . \tag{6.70}$$

If $\mathcal{N}_\alpha$ is a fractal set, then we expect the number of points in the set, $N_\alpha(\delta, R_g)$, to satisfy a scaling relation similar to equation (6.66):

$$N_\alpha(\delta, R_g) = \Delta\alpha\, \rho_\alpha(\delta, R_g) = \Delta\alpha\, b_\alpha \left(\frac{R_g}{\delta}\right)^{f(\alpha)} . \tag{6.71}$$

The number of points in the set is proportional to the range $\Delta\alpha$, so we have introduced the density, ρ_α, that is independent of this range. At this point let us again stress that the finite sets of points, which we necessarily consider when we discuss experimental results, only represent samples of the fractal sets $\mathcal{S}_\alpha$, which are defined only in the asymptotic limit of infinite systems or infinite resolution.

Equation (6.71) may in principle be used to determine the fractal dimension $f(\alpha)$ of the set that 'supports' the values of the measure specified by α, in the same way we determined the fractal dimension of the growth sites in the previous section. Unfortunately, we find that our models are too small to allow this direct approach to be used. We may nevertheless get an estimate for the $f(\alpha)$ curve using the measured values of μ_i. First we note that we may find the maximum value of $f(\alpha)$ by using the fact that the total number of growth sites is given by

$$N_I = \int d\alpha \, \rho(\alpha) \, . \tag{6.72}$$

From equation (6.66) and equation (6.71) we then find that

$$a \left(\frac{R_g}{\delta} \right)^{D_I} = \int d\alpha \, b_\alpha \left(\frac{R_g}{\delta} \right)^{f(\alpha)} \, . \tag{6.73}$$

This relation is valid for a large range in R_g and δ if the integrand has a sharp maximum at some value α_0. If this is the case, we may evaluate the integral by the method of steepest descent and we find that

$$f(\alpha_0) = D_I \, . \tag{6.74}$$

The amplitude a depends on the functional form of b_α and $f(\alpha)$, and cannot be determined in a general way. We also note that since the sets $\mathcal{N}_\alpha$ are subsets of the set of growth sites we have the relation

$$0 \leq f(\alpha) \leq D_I \, , \tag{6.75}$$

consistent with equation (6.74).

To obtain the $f(\alpha)$ curve we use the observed values of $\{\mu_i\}$ to make a histogram of $\rho(\alpha)$ and plot

$$f(\alpha) = \frac{\ln(\rho(\alpha)) - \ln b_0}{\ln(R_g/\delta)} \, , \tag{6.76}$$

as a function of α given by equation (6.69). The parameter b_0 represents the scale-independent part of the integral in equation (6.73), and is chosen so that the maximum value of $f(\alpha)$ is D_I, which occurs for $b_0 \sim 1.4$. The result of this analysis for three independent experiments is shown in figure 6.17. Note that b_α may in principle depend strongly on α. Therefore using b_0 instead of b_α in equation (6.76) in the analysis of the experiments implies that we obtain an *effective* exponent $f(\alpha)$. High-resolution experiments are required in order to determine the functional form of the amplitude b_α.

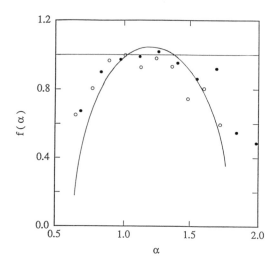

FIGURE 6.17: The $f(\alpha)$ curve for the fractal *new-growth measure* describing the dynamics of viscous fingering in porous models. Filled and open circles correspond to structures having $R_g/\delta \sim 170$ and 80 pixels, respectively. The curve is the $f(\alpha)$ obtained from $\tau(q)$ (Måløy et al., 1987b).

The Moments of the Measure

The fractal measure defined on the growth sites may also be analyzed by considering the moments of the observed measure μ_i, defined by

$$N(q, \delta, R_g) = \sum_{i=1}^{N_I} \mu_i^q = N(q) \left(\frac{R_g}{\delta} \right)^{\tau(q)} . \tag{6.77}$$

For $q = 0$, we find that $N(q = 0, R_g)$ equals the total number of islands N_I in the growth zone. We see from equation (6.66) and equation (6.77) that $\tau(q = 0) = D_I$, and $N(q = 0, R_g) = a$.

We determine $\tau(q)$ from fits of equation (6.77) to the experimental results, and determine $f(\alpha)$ and α from the relations (6.48). Using this transformation we converted the average $\tau(q)$ curve to the $f(\alpha)$ curve shown in figure 6.17.

Large values of α represent small μ_i. In our experiments we find it very difficult to determine very small values of μ because of the finite experimental resolution. This limits the accuracy of our results for large values of α. The same set of observations $\{\mu_i\}$, for three different experiments and thirteen different observations of the growth zone, have been used both in the present $\tau(q)$ analysis and in the direct analysis of the previous section.

The scatter of points found in the direct analysis gives a better representation of the limited experimental resolution than the smooth averaged curve determined from $\tau(q)$. Note that in both cases we used the observations of the scaling properties of N_I to fix unknown amplitudes. This in effect sets the maximum of $f(\alpha)$ to 1.

We emphasize that the harmonic measure $M_{\mathcal{H}}$ is different from the new-growth measure $M_{\mathcal{M}}$ representing the observed growth. The perimeter of DLA clusters is proportional to the cluster mass. All the perimeter sites have a probability of being hit by a random walker — they are the support of the harmonic measure. Therefore one expects that $f(\alpha)$ has a maximum value of 1.71, at an α value corresponding to $q = 0$. By contrast, the new-growth measure is supported by the new-growth sites $\mathcal{N}$. This set of sites is a fractal set with dimension $D_I \simeq 1$. Therefore one expects the maximum of $f(\alpha)$ to be D_I. The simulation estimating $M_{\mathcal{H}}$ by Amitrano et al. (1986) has a value of $f_{\max} \sim 1.5$, which is close to the fractal dimension 1.7 of the DLA aggregates.

Chapter 7

Percolation

Broadbent and Hammersley (1957) discussed the general situation of a *fluid* spreading *randomly* through a *medium*, where the abstract terms 'fluid' and 'medium' could be interpreted according to context.[1] The randomness may be of two quite different types. In the familiar diffusion processes the randomness is the random walks of the fluid particles — an example is the irregular thermal motion of molecules in a liquid. The other case in which the randomness is frozen into the medium itself, was christened a *percolation process* by Hammersley, since it behaves like coffee in a percolator.

Diffusion processes, such as the spreading of a solute in a solvent or electrons moving in a semiconductor, are well-understood processes. See chapter 9 for the interpretation of diffusion processes as random walks.

A diffusing particle may reach any position in the medium. Percolation processes are different. The most remarkable feature of percolation processes is the existence of a *percolation threshold*, below which the spreading process is confined to a *finite* region. An example discussed by Broadbent and Hammersley (1957) is the spread of a blight from tree to tree in an orchard where the trees are planted on the intersections of a square lattice. If the spacing between the trees is increased so that the probability for infecting a neighboring tree falls below a critical value, p_c, then the blight will not spread over the orchard. The percolation threshold for this problem is the probability $p_c = 0.59275$ for site percolation on a square lattice. Another example is the seepage of water, and perhaps radioactive waste, in the cracks and fractures of a rock formation. The question is whether the water is contained or will spread into other formations. Again a critical threshold for the concentration of cracks is expected. The value of the percolation threshold has to be determined by simulations. A similar

---

[1] For an account of the motivation for early work on percolation see Hammersley (1983).

problem of great practical interest is the spread of water displacing oil in porous rocks. Here the advancing fluid front may trap regions of oil, leading to *invasion percolation*, as discussed by Wilkinson and Willemsen (1983). The randomness encountered by the invading fluid now also depends on the dynamics through the formation of trapped regions. The concepts of percolation theory also apply to the propagation and interconnection of cracks and fractures in rocks and engineering materials.

There is no sharp distinction between percolation processes and diffusion in many applications. An important case is the diffusion of particles from a source. The resulting diffusion front has a geometrical structure that is closely related to the fractal geometry of percolation. This fact was first noted by Sapoval et al. (1985). We discuss this interesting case in section 7.9.

There is now a vast literature on percolation processes. A very nice introduction has been written by Stauffer (1985). Aharony (1986) and Aharony and Stauffer (1987) give a concise presentations and discuss several important applications. Earlier reviews by Shante and Kirkpatrick (1971), Kirkpatrick (1973), Stauffer (1979) and Essam (1980) give many essential details. Interesting contributions on recent developments in percolation theory and extensive bibliographies are found in the proceedings of various meetings — see for instance Deutscher et al. (1983), Pynn and Skjeltorp (1985), Englman and Jaeger (1986) and Pynn and Riste (1987).

The percolation problem is quite easily described and it leads to a wealth of very interesting fractal structures. We will mostly illustrate the concepts using two-dimensional percolation on a square lattice.

7.1   Site Percolation on a Quadratic Lattice

We occupy at random a fraction, $p = 0.5$, of the nodes of a square grid, i.e., a quadratic lattice, by objects as illustrated in figure 7.1. The objects illustrate pores in a matrix, and neighboring pores are connected by small capillary channels. A fluid injected into any given pore may only invade another pore that is directly connected to that pore through capillary channels or 'bonds.' The pores or 'sites' connected to the chosen center of injection form what is called a *cluster*. In figure 7.1 the largest cluster contains 46 sites, the next largest 29 sites, and so on. There are several clusters that contain only one pore. By studying figure 7.1 one quickly notes that none of the clusters spans the lattice. Thus, it is not possible to inject a fluid into a site on the left edge and have it come out somewhere on the right edge for the structure shown.

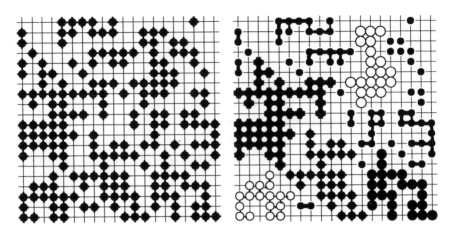

FIGURE 7.1: A quadratic lattice with one half of the nodes occupied by 'pores' is shown to the left. Connected regions, or 'clusters' are shown to the right. The largest clusters are distinguished by using different symbols for the pores. The lattice consists of $L \times L$ nodes with $L = 20$.

The effect of increasing the probability p of having an open pore on a site of the lattice is shown in figure 7.2 (a color version of this figure is found in the insert following the Contents). Clusters of various sizes are shown in different colors. The largest cluster (colored white), is seen to grow from a finite size at $p = 0.58$ into a very large cluster that contains a large fraction of the sites at $p = 0.62$. At $p = 0.6$ the largest cluster spans the lattice, connecting the left and right edges to the bottom edge. This cluster is called the *spanning cluster* or the *percolation cluster*. If the simulation is repeated one finds, of course, a new configuration of clusters. The spanning cluster first appears for $p = p_c \simeq 0.593$. Simulations on very large lattices have shown that the probability for having a spanning cluster vanishes as $L \to \infty$ when $p < p_c$. A finite fraction of the sites belong to the spanning cluster when $p > p_c$. The *critical probability* at which the spanning cluster first appears is $p_c = 0.59275 \pm 0.00003$, for site percolation on a quadratic lattice (Ziff, 1986).

The *percolation probability*, $P_\infty(p)$, is defined as the probability that a fluid injected at a site, chosen at random, will wet infinitely many pores. Note that the probability for having a pore at all at the site where fluid injection is attempted is p. The probability that the fluid wets infinitely many pores when it is injected into a pore that is known to belong to a cluster is $P_\infty(p)/p$. In practice we must consider finite systems consisting of N pores. For quadratic lattices $N = L^E$, where $E = 2$ is the Euclidian dimension of the space where the lattice is situated. In a simulation we

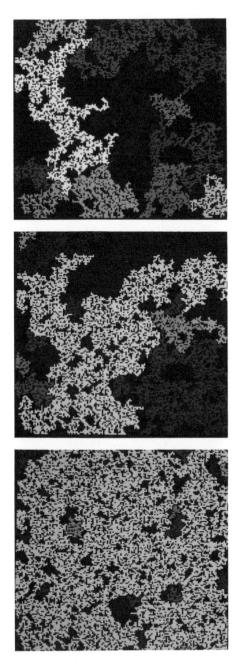

FIGURE 7.2: The effect of increasing the occupation probability p on a 160×160 quadratic lattice. From top to bottom we have $p = 0.58$, 0.6 and 0.62. In each of the three figures the largest cluster is shown in a light color. Smaller clusters are shown in darker shades. Unoccupied sites are black. A color version of this figure is included in the insert that follows the Contents.

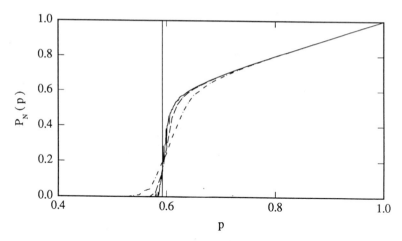

FIGURE 7.3: The probability, $P_N(p)$, of a site belonging to the largest cluster as a function of the probability p that a site is an open pore on an $L \times L$ square lattice. The full curve is obtained for $L = 450$, and the broken lines with $L = 200$ and 50. The vertical line is at $p = p_c = 0.59275$.

determine the number of sites, $M(L)$, that belong to the largest cluster on the $L \times L$ lattice and estimate $P_N(p)$ to be $M(L)/L^2$ averaged over many simulations similar to those shown in figure 7.2. The percolation probability is then given by

$$P_\infty(p) = \lim_{N \to \infty} P_N(p) \,. \tag{7.1}$$

For the quadratic lattice we have estimated $P_N(p)$ from simulations on $L \times L$ lattices with $L = 50$, 200 and 450 and we have we used 200, 100 and 10 independent samples respectively for these sizes. The resulting estimates of the percolation probability are shown in figure 7.3. For low concentrations, p, of pores we find that P_N is negligible. As p is increased the probability of belonging to the largest cluster increases drastically near $p_c = 0.593$, and then P_N increases almost linearly to 1 as $p \to 1$. The percolation transition at p_c becomes sharper as L is increased.

The *critical probability* is defined as the largest value of p for which $P_\infty = 0$. This may formally be written as

$$p_c = \sup \{p, \text{ such that } P_\infty(p) = 0 \} \,. \tag{7.2}$$

Thus, by definition, we have $P_\infty(p) = 0$ for $p \le p_c$. The percolation process undergoes a transition from a state of local connectedness to one where the connections extend indefinitely. Extensive simulations and theoretical work have shown that the percolation probability vanishes as a power-law *near* p_c:

$$P_\infty(p) \sim (p - p_c)^\beta \, , \quad \text{for } p > p_c \text{ , and } p \to p_c \, . \tag{7.3}$$

The exponent β is $5/36 = 0.1389\ldots$ for two-dimensional percolation processes, and $\beta \simeq 0.4$ for three-dimensional percolation. This is analogous to what happens at magnetic phase transitions, where the local order of magnetic moments increases in range as the temperature is lowered to the transition temperature, T_c, of the material. Below T_c, the magnetic moments are aligned, on the average, throughout the sample and we have a magnet. Many of the theoretical methods used to study phase transitions have been applied to percolation problems. Also the various critical exponents have been defined in analogy with the theory of phase transitions. The magnetization $m(T)$ vanishes as a power-law $(T_c - T)^\beta$ for many magnetic materials near the critical point. The exponent β is in the range 0.3–0.5 for most three-dimensional phase transitions. In the simulations one finds, of course, that P_N is finite even for $p < p_c$. It should be noted that in the simulations used to obtain figure 7.3, we find many examples where the largest cluster almost connects two opposite sides of the lattice for $p \simeq p_c$. These cases are not included in the estimates shown in figure 7.3.

7.2   The Infinite Cluster at p_c

How does the mass, or number of sites, $M(L)$, of the largest cluster grow with the size, L, of the lattice? For $p > p_c$, we expect that $M(L) \simeq P_N(p) \times L^2$, which tends to $P_\infty(p)L^2$ for $L \to \infty$, and $P_\infty(p)$ is simply the density of sites that belong to the percolating cluster. For $p < p_c$, on the other hand, we expect that $M(L)/L^2 \to 0$ as $L \to \infty$, since $P_\infty(p < p_c) = 0$. At p_c, one expects $M(L)$ to increase *almost* as L^2. The dependence of $M(L)$ on L has been studied extensively with the result[2]

$$M(L) \underset{L \to \infty}{\sim} \begin{cases} \ln L \, , & \text{for } p < p_c \, , \\ L^D \, , & \text{for } p = p_c \, , \\ L^E \, , & \text{for } p > p_c \, . \end{cases} \tag{7.4}$$

The mass of the percolating cluster is a *finite* fraction of the sites for $p > p_c$. Below p_c there is in general no spanning cluster. However, if one interprets

---

[2] Note that we use the sign $\sim$ both for asymptotically equal to and for asymptotically proportional to. This is customary in the percolation literature and makes it unnecessary to introduce symbols for the constants of proportionality (prefactor) that belong on the right-hand side of this equation.

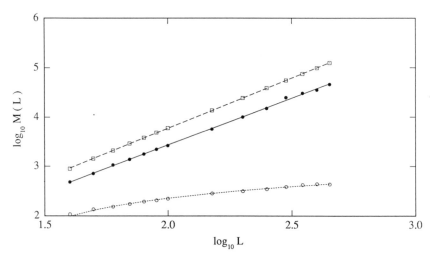

FIGURE 7.4: The mass of the largest cluster as a function of the linear dimension L of the quadratic lattice. Filled circles are for $p = p_c = 0.593$. The solid line is $M(L) = AL^D$ with $D = 1.89$. For $p = 0.65$ (open boxes) a fit (dashed line) gives $D = 2.03$. The results for $p = 0.5$, i.e., below p_c, have been fitted to the form $M(L) = A + B \ln L$, shown as the dotted line through the open circles.

$M(L)$ to be the size $s_{\max}$ of the *largest* cluster (e.g., Stauffer, 1985), then one finds that $M(L)$ increases only very weakly, i.e., logarithmically with L.

At the percolation threshold, $p = p_c$, the mass of the spanning cluster, which is also the largest cluster, increases with L as a power-law, L^D. Simulations on the square lattice give the results shown in figure 7.4. These results show that the percolation cluster at threshold is *fractal* with a cluster fractal dimension D; see equation (3.1). The fractal percolation cluster at threshold is often called the *incipient* percolation cluster. For the results shown in figure 7.4 we estimate $D \simeq 1.89 \pm 0.03$. Here the quoted errors are only statistical and represent the quality of the fit of the power-law to the results of the simulations shown in figure 7.4. Systematic errors are delicate to handle. When the percolation cluster on a finite lattice of side L is considered to be only a part of the incipient percolation cluster, then some of the sites not included in the percolation cluster on the scale L are really part of the incipient percolation cluster since they are connected to it by bonds outside the box under discussion. For $p > p_c$ we find that simulations on the quadratic lattice give a fractal dimension of $D = 2.03 \pm 0.01$ for the percolating cluster. Again the error is only statistical, and D represents the slope of the line through the points obtained by simulations at $p = 0.65$,

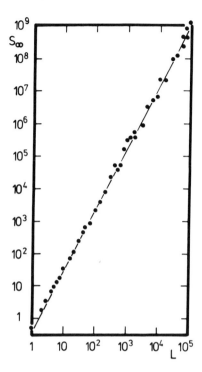

FIGURE 7.5: The size of the percolation cluster at the percolation threshold $p_c = \frac{1}{2}$ of the triangular lattice as a function of the linear dimension L of the lattice. The slope of this log-log plot for large L gives the fractal dimension $D = 91/48$ (Stauffer, 1985).

and shown in figure 7.4. The results for the largest cluster at $p = 0.5$, shown in figure 7.4, have been fitted to the form $M(L) = A + B \ln L$, with the result, $B = -426$ and $A = 327$. The dotted curve in figure 7.4 is the result of this fit. The results of the simulations shown in figure 7.4 are consistent with the asymptotic behavior given in equation (7.4).

Sykes and Essam (1964) have shown that the percolation threshold is $p_c = \frac{1}{2}$ *exactly* for site percolation on the triangular lattice. Therefore one may obtain very accurate results for the incipient percolation cluster by simulations on triangular lattices. The results obtained by Stauffer (1985) and shown in figure 7.5 give an estimate for the fractal dimension D consistent with the exact value $D = 91/48$. The numerical evidence suggests that this is the value for D for all two-dimensional lattice site percolation problems.

We conclude that the incipient percolation cluster has a *fractal struc-*

ture and the mass of this cluster increases *on the average* with L as

$$M(L) \sim \overline{A}L^D , \quad D = 91/48 = 1.895\ldots , \qquad (7.5)$$

for site percolation on two-dimensional lattices. The average is over many realizations of the incipient percolation cluster. The amplitude $\overline{A}$ is the effective amplitude estimated from finite-size samples. This scaling law for the mass of the incipient percolation cluster is only valid *asymptotically*, for large L. At realistic values of L this scaling relation should be modified by *correction terms* (e.g., Margolina et al., 1984; Aharony, 1986):

$$M(L) = AL^D + A_1 L^{D_1} + A_2 L^{D_2} + \ldots , \qquad (7.6)$$

with $D > D_1 > D_2$. It is difficult to determine the correction terms by direct simulations. Recently Aharony et al. (1985) proposed a novel transfer matrix method, which makes this task easier. Typically one finds that $D_1 \simeq D - 1$, for two-dimensional problems (Grossman and Aharony, 1986).

Note that the Mandelbrot-Given curve (see figure 2.13) has a fractal dimension of $D = 1.892\ldots$ and is a good model for the percolation cluster.

7.3   Self-Similarity of Percolation Clusters

The incipient infinite cluster (the percolation cluster) is *statistically self-similar*. Consider the percolation cluster at $p = 0.6$ shown in the middle panel of figure 7.2. If we view this cluster at a lower resolution then the details will become blurred, but it looks similar. The overall structure of the cluster, for instance the fact that there are holes of all possible sizes in the cluster, remains. This self-similarity is intimately linked to the fractal structure of the incipient percolation cluster, and may be made quantitative by *real space renormalization*. This renormalization is best illustrated using percolation on the triangular lattice, where $p_c = 1/2$.

Consider the percolation clusters on a triangular lattice with $p = p_c$ shown in figure 7.6. For the triangular lattice we may change the scale of the lattice by a factor $b = \sqrt{3}$ by grouping occupied sites, i.e., open pores, as shown in figure 7.7. Basic cells of $b^2 = 3$ sites are replaced by single new sites, which are considered to be open pores (occupied sites) if a majority of the sites in the cell are occupied. This rescaling ensures that a cluster of two or three open pores becomes an open pore in the coarse grained lattice shown in figure 7.7b (see Young and Stinchcombe, 1975; Reynolds et al., 1977).

The result of this coarse-graining is a new lattice with a *new* concentration p' of occupied sites. For the simple example shown in figure 7.7a

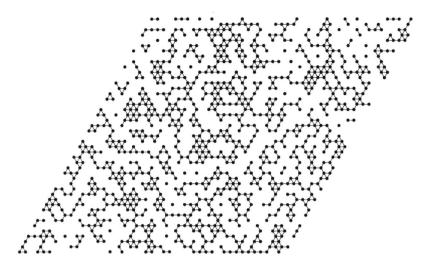

FIGURE 7.6: Site percolation clusters on the triangular lattice at the threshold $p_c = 1/2$.

we note that the probability of finding a block of three occupied sites is p^3, since the sites are independently occupied with a probability p. The probability of having two sites occupied is $3p^2(1-p)$, since there are three orientations of the empty site. It follows that the new concentration p' is given by (Reynolds et al., 1977)

$$p' = p^3 + 3p^2(1-p) . \tag{7.7}$$

One may coarse-grain the new triangular lattice again, and again. In each iteration the new concentration of occupied sites is given in terms of the old concentration by equation (7.7). We have plotted in figure 7.8 the change in concentration, $p' - p$, caused by one iteration of the renormalization procedure as a function of the concentration of occupied sites, p, before the rescaling. Note that $p = p_c = 0.5$ solves the iteration equation (7.7), so that p_c is a fixed point for the rescaling. If one starts out with a lattice with $0 < p < p_c$, then $p' - p < 0$, and the new concentration of occupied sites is less than the original concentration. Therefore, if we start with a large triangular lattice and apply the renormalization transformation many times, we eventually get a lattice of empty sites. If, on the other hand, we start with $1 > p > p_c$ then the concentration increases in each iteration and we will end up with a coarse-grained lattice with all sites filled, i.e., by open pores. At the *critical point*, $p = p_c$, we find that the renormalization does *not* change the concentration of occupied sites, and the coarse-grained lattice is also at the percolation threshold. The critical point is a *fixed-point* of the renormalization transformation. Of course, there are in addition the two trivial fixed points $p = 0$ and $p = 1$.

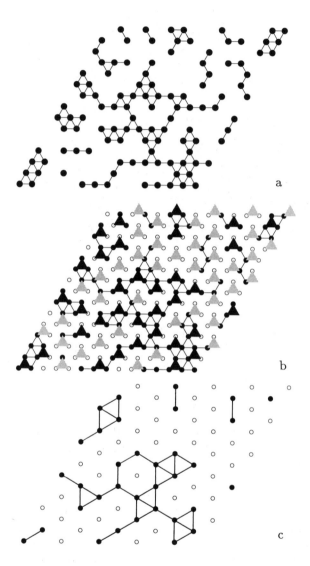

FIGURE 7.7: (a) Clusters at p_c on a triangular lattice. (b) Coarse-graining of the triangular lattice by a factor $b = \sqrt{3}$. Occupied sites are shown as filled circles and open sites are shown as open circles. Groups of three sites are blocked as indicated by the shaded triangles. If a majority of the sites belonging to a triangle are occupied then the triangle is shaded dark. (c) The coarse-grained percolation lattice where filled sites correspond to dark triangles in (b).

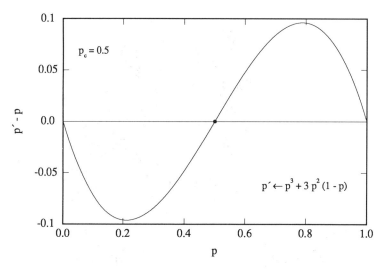

FIGURE 7.8: The change in the concentration of occupied sites, $p' - p$, as a function of the concentration of occupied sites, p, on the triangular lattice before a rescaling of the lattice by a factor of $b = \sqrt{3}$.

The effect of the renormalization procedure illustrated in figure 7.7 on a large triangular lattice at the percolation threshold is shown in figure 7.9. Qualitatively, the pattern of occupied sites obtained by using the renormalization transformation twice (see figure 7.9b) cannot be distinguished from a piece of the original lattice. Also the percolation cluster of the scaled lattice, shown in figure 7.9c, is qualitatively the same as the original percolation cluster. It is important to note that the scaling transformation does not change the site occupation probability p, and therefore a system at p_c remains at threshold even after the scale transformation. It is impossible to tell from the pictures at what level of coarse-graining, or magnification, the pictures were taken. This aspect of self-similarity is illustrated by inserting a scaled version of the lattice into the original lattice at some position. Statistical self-similarity implies that the resulting percolation cluster is an equally probable realization of the process that generates percolation clusters.

The statistical self-similarity of the incipient percolation cluster may be used to obtain a quantitative statement for the mass of the cluster. The scaling law (7.5) for the mass $M(L)$ of the cluster must also apply for the mass of the cluster obtained after a scaling by a factor of b. But the linear size of the scaled lattice is only L/b, and therefore we find that the number of sites, $M(L/b)$, must be $\overline{A}(L/b)^D$. It follows from equation (7.5) that the

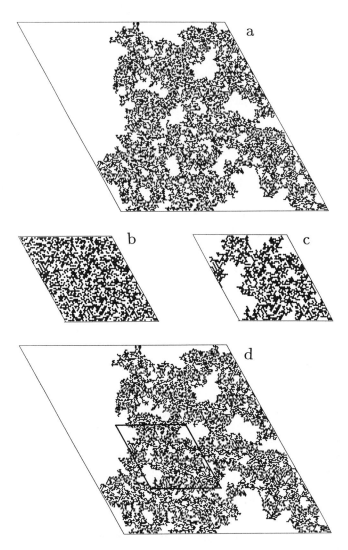

FIGURE 7.9: Self-similarity and the effect of coarse-graining on the trian-gular lattice at the percolation threshold. The scale is changed by a factor $b = 3$, using the renormalization procedure illustrated in figure 7.7 twice. (a) The percolation cluster at threshold on the original 180 × 180 lattice. (b) The occupied sites obtained after a scaling of the original lattice in (a) by a factor $b = 3$. (c) The percolation cluster of the scaled lattice in (b). (d) The scaled lattice is inserted into the original lattice in the region marked by a frame. The figure shows the resulting percolation cluster.

cluster masses at the two scales are related by the scaling relation

$$M(L) = b^D M(L/b) . \tag{7.8}$$

This relation is valid only asymptotically in the limit of large L and L/b. However, it is valid for *all* values of the scale factor b consistent with this constraint. Conversely one finds, of course, that the scaling equation (7.8) implies a power-law behavior. Since the left-hand side of equation (7.8) is independent of b it follows that $M(L)$ must be a *homogeneous function*, and the only possible form for $M(L)$ is the power-law $M(L) \sim L^D$. The fractal geometry of the incipient percolation cluster and the statistical self-similarity are related and expressed quantitatively by equation (7.8).

7.4 Finite Clusters at Percolation

At percolation there is a wide distribution of cluster sizes (see figure 7.2). As the occupation probability is decreased below p_c the clusters gradually decrease in size. Above p_c, there are clusters of various sizes in the holes of the percolating cluster. The number of sites, s, in a cluster and its linear extension have characteristic distributions. The percolation threshold is characterized by a cluster size distribution that has no typical size, i.e., it must be a power-law distribution. To make this statement more precise let us introduce the radius of gyration, $R_g(s)$, of a cluster consisting of s sites:

$$R_g^2(s) = \frac{1}{2s^2} \sum_{i,j} (\vec{r}_i - \vec{r}_j)^2 . \tag{7.9}$$

The radius of gyration is simply the root mean square radius of the cluster measured from its center of gravity. Consider the finite cluster illustrated in figure 7.10. When the finite cluster (at p_c) is analyzed inside a box with side $L \leq 2R_g(s)$, then it appears to be a part of the incipient percolation cluster spanning the box, and one finds that $M_s(L) \sim L^D$, as before. However, when the box size is increased beyond $2R_g$, one begins to see the edges of the cluster. For sufficiently large L, the whole cluster fits inside a box of size L_s, and there is no doubt that the cluster is finite since the mass of the cluster no longer increases with L. These considerations may be summarized as follows: The mass, $M_s(L)$, inside a box of size L on a cluster that consists of s sites is given by

$$M_s(L) = L^D f(L/R_g) \rightarrow \begin{cases} \overline{A}(L/R_g)^D , & \text{for} \quad L \ll R_g(s) , \\ s , & \text{for} \quad L \gg R_g(s) . \end{cases} \tag{7.10}$$

Here the crossover function $f(x)$ simply tends to the constant amplitude $\overline{A}$ in equation (7.5) as $x = L/R_g(s) \rightarrow 0$. However, since $M_s(L)$ must

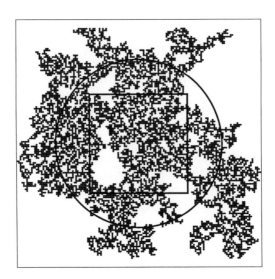

FIGURE 7.10: A finite cluster on the square lattice at p_c. The radius of the circle is the radius of gyration, $R_g(s) = 51$, of a cluster containing 6700 sites. The box indicated in the figure has a side $L = 60$. The side of the smallest box that contains the cluster is $L_s = 150$.

become independent of L for $x \gg 1$, we may conclude that $f(x) \sim x^{-D}$, so that the term L^D in front of f is canceled in equation (7.10). We therefore find the following relation between the radius of gyration and the number of sites in the cluster:

$$s = M_s(L \gg R_g) \sim L^D (L/R_g)^{-D} \sim R_g^D . \qquad (7.11)$$

The relation $s \sim R_g(s)^D$ has been confirmed by many computer simulations. In figure 7.11 we show the results obtained by Grossman and Aharony (1986) and Aharony (1986) for the dependence of s on L_s, where L_s is the dimension of the smallest box that will hold the cluster. The clusters have no intrinsic length scale independent of cluster size, and therefore one also expects $M(L_s)$ to scale as L_s^D. This is brought out clearly in figure 7.11.   The extrapolation of the 'effective' fractal dimension estimated from a part of the s versus L_s curve by $D_{\text{eff}} = \partial \ln s / \partial \ln L_s$, shown in the insert, gives the expected value for the fractal dimension $D = 1.89 \pm 0.01 \simeq 91/48$, in the limit of large clusters. The fact that D_{eff}, can be fitted to a straight line in $1/L_s$ indicates that the leading correction D_1 in equation (7.6) is given by $D_1 = D - 1$. The error bars indicate the scatter of the values of s within 'windows' of linear size L_s. The existence of this scatter again emphasizes the fact that the power law $s \sim L_s^D$ applies only to the *average*. The fact that the error bars have fixed length on

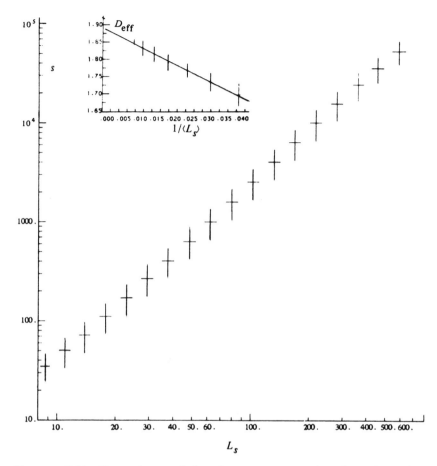

FIGURE 7.11: Dependence of the cluster masses s (number of sites) on their linear size L_s, for the square lattice at $p_c = 0.5927$. The error bars indicate one standard deviation around the average. The insert shows the extrapolation of the local slope $D_{\text{eff}} = \partial \ln s / \partial \ln L_s$ to $s \to \infty$, where $D = 1.89 \pm 0.01$ (Aharony, 1986).

logarithmic scales leads to the conclusion that the fluctuations in s for a given L_s are given by

$$\langle s^2 \rangle - \langle s \rangle^2 \sim L_s^{2D} . \tag{7.12}$$

Such fluctuations are referred to as *lacunarity*; see Mandelbrot (1982) and Aharony (1986).

7.5   The Cluster Size Distribution at p_c

Let us now consider the distribution of sizes for the clusters that are found at the percolation threshold and are illustrated in figure 7.2. Let n_s be the average number *per site* of clusters containing s sites, that is, the number of clusters of size s that is expected on an $L \times L$ lattice is $n_s L^2$. Thus if we select a site at random it has a probability proportional to $s n_s$ of being on a cluster of size s, since there are s ways to 'hit' such a cluster. The normalized probability per site, c_s, that a site chosen at random is part of a cluster containing s sites is then given by

$$c_s = \frac{s\, n_s}{\sum_s s\, n_s}\ . \tag{7.13}$$

The normalization term $\sum_s s\, n_s$ is simply the fraction of sites that belong to *finite* clusters. The average cluster mass is then

$$\langle s \rangle = \sum_{s=1}^{L^D} s \cdot s n_s(L) \bigg/ \sum_s s\, n_s\ . \tag{7.14}$$

For a simulation on a finite lattice of side L, we must find that n_s depends on L, and we have therefore written $n_s(L)$ in equation (7.14). The sum extends over all the cluster sizes from the smallest, $s = 1$, to the largest, which is of the order $s_{\max} \sim L^D$. We may find the asymptotic dependence of $n_s(L)$ using scaling arguments.

Consider an $L \times L$ lattice at the percolation threshold. A scale transformation by a factor b, gives as a result an $(L/b) \times (L/b)$ lattice that is also at the percolation threshold. The probability that a random site on the $L \times L$ lattice is part of a cluster of size s is $s n_s(L)$. The scale transformation reduces the radius of gyration of a cluster n_s be the average number *per site* of containing s sites from $R_g(s)$ to $R_g(s') = R_g(s)/b$. It follows from equation (7.11) that a cluster containing s sites is mapped into a cluster having only $s' = s/b^D$ sites. We may therefore express the probability $s n_s(L)$ in terms of the probability $s' n_{s'}(L/b)$ that a site on the $(L/b) \times (L/b)$ lattice belongs to a cluster of a size s' that maps into a cluster of size s:

$$s n_s = b^{-2} s' n_{s'}\ , \quad \text{with } s' = s/b^D\ . \tag{7.15}$$

Here the factor b^{-2} on the right-hand side is due to the fact that for each site on the $(L/b) \times (L/b)$ lattice there are b^2 sites on the $L \times L$ lattice. The right-hand side must be divided by b^2 since we are considering probabilities *per site*.

The cluster size distribution depends on L only through the variable $s/s_{\max} \sim 2/L^D$, because the largest cluster in an $L \times L$ lattice is limited to about $s_{max} \sim L^D$ sites. Therefore we use the following form for $n_s(L)$:

$$n_s(L) = s^{-\tau} g(s/L^D) . \tag{7.16}$$

Here $g(x)$ is a crossover function that must approach a constant value as $s/L^D \to 0$, since the cluster size distribution must become independent of the size of the lattice in the limit $L \to \infty$. The power-law dependence in equation (7.16) is required since n_s should satisfy equation (7.15). Inserting equation (7.16) into equation (7.15) we get the following equation:

$$
\begin{aligned}
s^{1-\tau} g(s/L^D) &= b^{-2} \left(s/b^D\right)^{1-\tau} g \left[(s/b^D)/(L/b)^D\right] \\
&= b^{-(2+D-\tau D)} s^{1-\tau} g(s/L^D) .
\end{aligned}
\tag{7.17}
$$

From this equation we conclude that $2 + D - \tau D = 0$, and the cluster size distribution at the percolation threshold is a power-law,

$$n_s \sim s^{-\tau} , \quad \text{with } \tau = (E+D)/D . \tag{7.18}$$

The exponent τ depends on the Euclidean dimension, $E = 2$, of the lattice. We write E in the expression for τ since equation (7.18) holds also for site percolation on hypercubic lattices with $E < 6$. For $E \geq 6$, special considerations apply; see Aharony (1986) for a discussion.

We may rewrite equation (7.16) as follows:

$$
\begin{aligned}
n_s(L) = s^{-\tau} g(s/L^D) &= L^{-D\tau} n \left(s/L^D\right) \\
&= L^{-E-D} n(s/L^D) .
\end{aligned}
\tag{7.19}
$$

Here the new crossover function $n(x)$ is given by $n(x) = x^{-\tau} g(x)$. This form for n_s is convenient when we want to estimate how various averages vary with the size L of the lattice. For instance, the average cluster size given by equation (7.12) is evaluated as follows:

$$
\begin{aligned}
\langle s \rangle_L &= \sum_{s=1}^{L^D} s^2 n_s(L) \Big/ \sum_s s\, n_s \\
&\simeq L^{2D-E} \int_{1/L^D}^{1} d\frac{s}{L^D} \left(\frac{s}{L^D}\right)^2 n(s/L^D) \\
&\sim L^{2D-E} .
\end{aligned}
\tag{7.20}
$$

The normalization term $\sum_s s\, n_s = p_c$, since it is the fraction of sites that belong to finite clusters. The last relation results because the integral approximation to the sum depends only on the variable $x = s/L^D$, and is convergent at the limits. Therefore the integral becomes independent of L

in the limit of large L. This result shows that the average cluster size at the percolation threshold, $\langle s \rangle_L \sim L^{2D-E}$, grows more slowly with L than the size of the largest cluster $s_{\max} \sim L^D$.

In a similar way we may find the average radius of gyration of the finite clusters at the percolation threshold

$$\langle R_g^2(s) \rangle_L = \frac{\sum_s R_g^2(s) s\, n_s}{\sum_s s\, n_s} \sim L^{2-(E-D)} . \tag{7.21}$$

Note that this result means that *not* every length scale one may define in the percolation problem scales as the lattice size L, since $\langle R_g^2(s) \rangle_L$ is *not* proportional to L^2; see Stauffer (1979).

The arguments presented are typical for the scaling arguments used in percolation theory and in the theory of critical behavior at second-order phase transitions. The essential aspect of the arguments presented lies in the *self-similar* structure of the percolation process at threshold. This self-similarity leads to power-law dependencies of the various quantities considered. However, not every power-law exponent is a fractal dimension. Many of the exponents that arise may be expressed in terms of the fractal dimension and the spatial dimension of the lattice under consideration. The equations (7.20) and (7.21) are examples of *finite size scaling* used widely to identify critical exponents from computer simulations at p_c.

7.6   The Correlation Length ξ

The clusters both at p_c and away from p_c are characterized by the number of sites, s, in the cluster and by the radius of gyration, $R_g(s)$, of the cluster. Over what distances are open pores (sites) connected? The *connectedness length* ξ is defined as the *average* root mean square distance between occupied sites that belong to the *same* and *finite* cluster. This connectedness length is also called the *correlation length*. The connectedness length is then the square root of $R_g^2(s)$ averaged over the cluster size distribution.

To find an expression for ξ, consider a site on a cluster consisting of s sites. The site is connected to $s - 1$ other sites, and the average square distance to those sites is $R_g^2(s)$. The probability that a site belongs to a cluster of size s is $s n_s$. The connectedness length is consequently given by

$$\xi^2 = \frac{2 \sum_s R_g^2(s) s^2\, n_s(p)}{\sum_s s^2\, n_s(p)} . \tag{7.22}$$

Here we have written s instead of $s-1$ for the number of sites to which a given sites is connected in order to simplify the expression. The cluster size distribution, $n_s(p)$, is now a function of p and is the average number per site of clusters of the *finite* size s.

The equation (7.21) shows that the radius of gyration diverges with the system size at p_c and we conclude that $\xi = \infty$ for $p = p_c$. Near p_c one finds that the pair connectedness length diverges as a power-law:

$$\xi(p) \sim |p - p_c|^{-\nu} . \qquad (7.23)$$

The exponent $\nu = 4/3$ for two-dimensional percolation. In fact one may understand how this power-law arises and obtain a very good approximate expression for ν based on the renormalization of the triangular lattice discussed in section 7.3. When the triangular lattice at a concentration p is coarse-grained by a factor of $b = \sqrt{3}$, then the new lattice has the concentration p' given by equation (7.7). The new correlation length ξ' is given by the relation

$$\xi' = \xi(p') = \xi(p)/b . \qquad (7.24)$$

Near the fixed point p_c the renormalization transformation (7.7) is linear in $(p-p_c)$ as shown in figure 7.8. We find by an expansion of equation (7.7) in $(p - p_c)$ around $p_c = 1/2$ that p' may be written as

$$p' = p_c + \Lambda (p - p_c) , \quad \text{for} \quad |p - p_c| \ll p_c , \qquad (7.25)$$

with $\Lambda = 3/2$. Using this result in the equation (7.24) we find that ξ is a *homogeneous* function of $p - p_c$:

$$\xi(\Lambda|p - p_c|) = b^{-1}\xi(|p - p_c|) \qquad (7.26)$$

Again, the power-law equation (7.23) is the only form that satisfies this relation, and inserting equation (7.23) here we find that

$$(\Lambda|p - p_c|)^{-\nu} = b^{-1}|p - p_c|^{-\nu} . \qquad (7.27)$$

From this equation it follows that the exponent ν, which controls the divergence of the correlation length at p_c, is given by

$$\nu = \ln b / \ln \Lambda = \ln \sqrt{3} / \ln(3/2) \simeq 1.355 , \qquad (7.28)$$

which is an excellent approximation to the exact two-dimensional value $\nu = 4/3$, found by den Nijs (1979). This derivation is a simple example of the general method for calculating critical exponents by the renormalization transformation. Note that the rescaling transformation used is approximate since the connectivity of the clusters may change by the transformation. A more detailed discussion is given by Aharony (1986).

We may find the behavior of the cluster size distribution, $n_s(p)$, by scaling arguments. Consider clusters that have radii of gyration $R_g(s) < \xi$. The number of sites in such clusters must satisfy the relation $s < \xi^D$, since the radius of gyration scales as $s \sim R_g^D$. On length scales less than ξ we have no way of telling that we are not at p_c, and we conclude that the size distribution on these scales must be given by equation (7.19), but now with the box length L replaced by ξ. Therefore we find that

$$n_s(p) = \begin{cases} \xi^{-E-D} \, \mathcal{N}(s/\xi^D) \, , & \text{for} \quad s \ll \xi^D \, . \\ 0 \, , & \text{for} \quad s \gg \xi^D \, . \end{cases} \tag{7.29}$$

The crossover function $\mathcal{N}(x)$ decays very quickly for $x \gg 1$, and behaves as $x^{-\tau}$ for $x \ll 1$ just as the crossover function $n(x)$ in equation (7.19). We may use this form of the cluster size distribution to calculate the average cluster size and we find that

$$\langle s \rangle_\xi = \frac{\sum_s s^2 \, n_s(p)}{\sum_s s \, n_s} \sim \xi^{2D-E} \, , \tag{7.30}$$

by the same method we used to obtain equation (7.20).

The probability (per site) of a site belonging to the *infinite* percolating cluster is $P_\infty(p)$ for $p \geq p_c$. Any given site is occupied with a probability p, and belongs to one of the finite clusters with probability $\sum_s s \, n_s(p)$. Sites that do not belong to the finite clusters must belong to the infinite cluster and therefore we find the relation

$$P_\infty(p) = p - \sum_s s \, n_s(p) \, . \tag{7.31}$$

Using the scaling form of the cluster size distribution one finds that the percolation probability scales as[3]

$$P_\infty \sim (p - p_c)^\beta \sim \xi^{D-E} \, . \tag{7.32}$$

From this relation we obtain an expression for the exponent β for the percolation probability:

$$\beta = (E - D)\nu = \tfrac{5}{36} \text{ for } E = 2 \text{ and } 0.4 \text{ for } E = 3 \, . \tag{7.33}$$

The number of sites in the largest cluster, $M(L)$, depends now on how close one is to p_c. For boxes of size $L \ll \xi$, the largest cluster spans the box and $M(L) \sim L^D$. However, for $L \gg \xi$ the sizes of the finite clusters

---

[3]Note that one has to be careful with the limits of the integral representation of the sum in this case. See Stauffer (1985,1986) for a discussion.

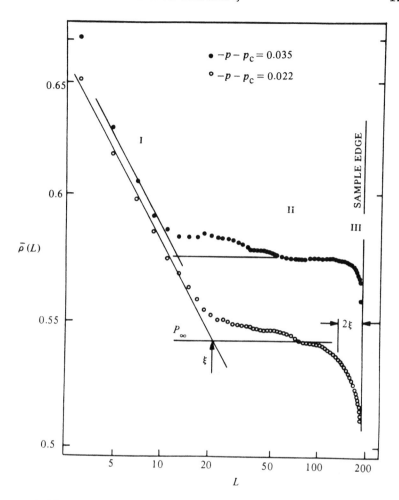

FIGURE 7.12: Density of sites on the largest cluster of a square lattice at $p - p_c = 0.35$ (filled circles) and $p - p_c = 0.022$ (empty circles) within a box of size L around an occupied site. The slope for $L < \xi$ is $D - 2$ with $D \simeq 1.9$, and the plateau for $L > \xi$ is $P_\infty(p)$ (Kapitulnik et al., 1983).

are cut off at ξ, and $M(L)$ should cross over to $M(L) = \bar{\rho} L^E$, with the average density $\bar{\rho}$ given by $\bar{\rho} = P_\infty \sim \xi^{D-E}$. This crossover is controlled by a scaling function $m(x)$, so that

$$M(p, L) = L^D \, m(L/\xi) \,, \quad \text{with} \quad \xi \sim |p - p_c|^{-\nu} \,. \qquad (7.34)$$

Numerical simulations shown in figure 7.12 show this crossover very clearly.

We see from the discussion in this section that the connectedness length, or correlation length, ξ diverges at the critical concentration. The divergence is controlled by a power-law with an exponent that may be calculated quite accurately by the renormalization transformation. In practice it is convenient to express the various crossover behaviors and the power-laws near the critical concentration in terms of ξ. Also, one finds that the various exponents found near the percolation threshold are related by scaling laws which express the exponents in terms of the fractal dimension, D, of the percolation cluster, the spatial dimension $E = 2$ and the exponent ν for the correlation length.

7.7   The Percolation Cluster Backbone

We have discussed percolation in terms of a 'fluid' wetting 'pores' when injected from a single site. This discussion presupposes the pores are empty so that a fluid may actually enter each pore. A realization of this process is the injection of mercury into a porous material that has been evacuated before the injection process.

Consider pores that form a connected network and are filled by an incompressible fluid (oil). Another fluid (water) that is injected can only displace the oil on the *backbone* of the percolation cluster. The parts of the percolation cluster that are connected to the backbone by only a single site are called *dangling ends*. It is sufficient to remove a single site, i.e., cut a single *dangling bond*, to separate the dangling end from the backbone. The driving fluid (water) cannot enter the dangling ends since the trapped oil has no escape route.

The backbone consists of all the sites visited by all possible self-avoiding walks from the injection site(s) to the exit site(s). A dangling end may not be visited by a self-avoiding random 'walker' since it must retrace at least one step to escape the region connected to the backbone by a single site.

A realization of a percolation cluster and a backbone is shown in figure 7.13 for site percolation on the quadratic lattice at threshold. The backbone connects a single site at the center of the 147 × 147 lattice to the sites at the edges of the square lattice. The number of pores on the percolation cluster is 6261, whereas the backbone contains only 3341 sites.

We have made a physical model (Oxaal et al., 1987) of the percolation cluster shown in figure 7.13. The model was molded using epoxy and has cylindrical pores 1.1 mm in diameter and 0.7 mm high. The pores are connected by 0.7 mm wide channels. The model was filled with high-viscosity colored glycerol. In a typical displacement experiment air is injected at the

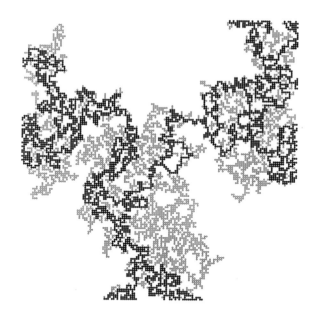

FIGURE 7.13: The backbone (filled circles) and the percolation cluster from a simulation on a 147 × 147 quadratic lattice at $p_c = 0.593$ (Oxaal et al., 1987).

site in the center of the model, thereby displacing the glycerol, which flows out of the model at the rim. The experimental result shown in figure 7.14 very clearly illustrates the fact that displacement processes take place only on the backbone.

The viscous displacement process on the fractal percolation cluster can be modelled numerically by solving the flow equation (4.3) with the appropriate boundary conditions (Murat and Aharony, 1986; Oxaal et al., 1987). The results of simulations for flows at high capillary numbers on the percolation cluster used in the experiment are also shown in figure 7.14.

The agreement between the experiment and the simulation is very good. In fact 70–80% of the sites invaded by air are common to the experiment and the simulation, at any time in the invasion process. Separate simulations also overlap each other to about 75%. This agreement shows that fluid displacement at the percolation threshold is almost entirely determined by *geometrical* effects since the numerical simulation does not take into account such factors as interfacial tension and wetting properties that are known to influence ordinary two-phase flow in porous media.

The backbone of the percolation cluster depends on what is considered to be the injection site(s) and exit site(s). As an example consider the

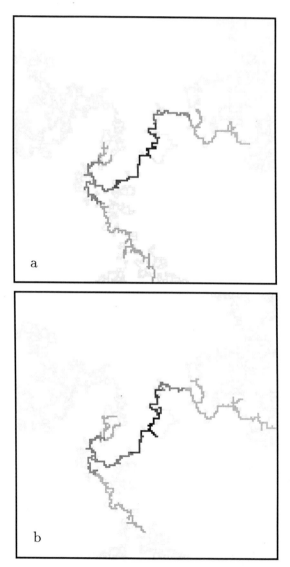

FIGURE 7.14: (a) Air displacing glycerol at a high capillary number on the percolation cluster shown in figure 7.13. (b) The results of numerical simulation of fluid displacement on the same percolation cluster. The different shades of gray represent pores invaded by air observed at successive time steps. The number of pores invaded by air is 30 (black), 86, 213 and finally at breakthrough 605 light gray, for both the experiment and the simulation. The backbone is shown in very light gray color. A color version of this figure is included in the insert that follows the Contents. (Oxaal et al., 1987).

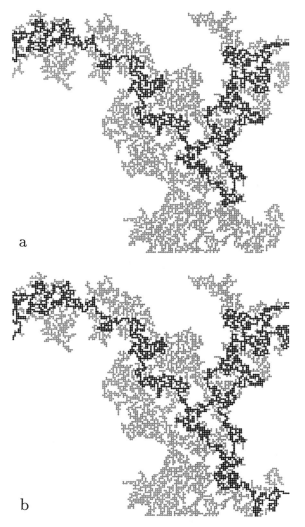

a

b

FIGURE 7.15: Backbones of the percolation cluster on the quadratic lattice at p_c shown in figure 7.2b. (a) The backbone connecting a single site on the left-hand edge to a single site on the right-hand edge of the 160 × 160 lattice. (b) The backbone connecting all the sites on the left-hand edge to all the sites on the right-hand edge.

percolation cluster at p_c shown in figure 7.2b. The backbone connecting a single site on the left edge to a single site on the right edge is shown in figure 7.15a whereas the backbone connecting all the sites on the left edge to sites on the right edge is shown in figure 7.15b.

The backbone sites form a subset of the sites on the percolation cluster and every site on the percolation cluster is also part of at least one backbone. Since the percolation cluster is fractal with a fractal dimension of $D = 1.89$, it follows that backbones on the percolation cluster are fractal, and have fractal dimensions $D_B \leq D$. Extensive numerical simulations (at p_c) have shown that the mass $M_B(L)$ of the backbone connecting the edges of a box with side L is given by

$$M_B(L) \sim L^{D_B} , \quad \text{with} \quad D_B \simeq 1.61 . \tag{7.35}$$

The estimates for the backbone fractal dimension have been in the range $D_B \simeq 1.62\pm0.02$ (Herrmann and Stanley, 1984), and have recently been estimated to be $D_B = 1.61\pm0.02$ by Laidlaw et al. (1987). The Mandelbrot-Given curve, shown in figure 2.14, is a reasonable model for the percolation cluster backbone and the fractal dimension $D_B = 1.63\ldots$ of that curve is quite close to the backbone fractal dimension.

The viscous displacement process on the percolation cluster selects only a *subset* of the sites in the backbone. The subset selected depends on the capillary number. Experiments and numerical simulations (Oxaal et al., 1987) indicate that the fractal dimension of the viscous fingering structure is $D \sim 1.3$ at high Ca, and 1.5 at low Ca.

In general we obtain various fractal dimensions in studying *physical* phenomena occurring on fractals. This is due to the fact that by specifying physical processes on the supporting fractal geometry we in effect specify *measures* on the underlying geometry. The study of physical phenomena on fractals therefore leads in a natural way to the multifractals discussed in the previous chapter. Current distributions and resistance fluctuations in fractal networks of (nonlinear) resistors give rise to infinite sets of exponents or multifractals (see de Arcangelis et al., 1985; Rammal et al., 1985; Blumenfeld et al., 1986, 1987; Aharony, 1987).

The backbone has many geometrical features that also are fractals. Consider the two points connected by the backbone shown in figure 7.15a. The shortest path ℓ_{min} between these points (measured by counting the number of sites one has to visit on the path) is found to scale with the box size L, i.e., the Euclidean distance between the points, as (see e.g. Havlin and Nossal, 1984)

$$\ell_{min} \sim L^{D_{min}} , \quad \text{with} \quad D_{min} = 1.15 \pm 0.02 . \tag{7.36}$$

When one studies pictures of backbones such as the ones shown in figure 7.15, it becomes clear that the backbone consists of '*blobs*' connected by '*links*' (Skal and Shklovskii, 1975; de Gennes, 1976; Stanley, 1977). The links, also called *red bonds* by Stanley (1977), have the property that if

they are cut then the backbone is separated into two parts and a fluid can
no longer flow from the injection site to the extraction site. The blobs are
multiply connected, so that cutting a bond, i.e., removing a site, will not
interrupt the flow. Stanley calls the bonds connecting sites in the blobs
blue bonds. The reason for the color scheme is that in an electrical analog,
where current flows through the percolation cluster from a single contact
(injection) site on one end of the cluster to another contact site at the
other end of the cluster, all the current must pass through the red bonds
and they become hot. In the blobs the current may spread over many
bonds and they remain relatively cool. The set consisting of the red bonds
forms a subset of the sites on the backbone and is in fact a fractal set of
points (Pike and Stanley, 1981). The number of red bonds diverges as the
separation L between the two sites on the ends of the backbone increases
according to a power-law:

$$N_{\text{red}} \sim L^{D_{\text{red}}} \ , \quad \text{with} \quad D_{\text{red}} = 1/\nu = {}^3\!/_4 \ . \tag{7.37}$$

The relation $D_{\text{red}} = 1/\nu$, between the fractal dimension of the red bonds
and the exponent ν that controls the divergence of the correlation length ξ
at p_c, was also shown to hold rigorously in higher dimensions by Coniglio
(1981, 1982). Almost all the mass of the backbone is in the blobs, since the
fractal dimension of the red bonds is much less than that of the backbone.
The fractal dimension of the sites that belong to blobs therefore equals that
of the backbone. The Mandelbrot-Given curves (see figures 2.13 and 2.14)
have many singly connected (red) bonds. The fractal dimension of these
bonds is $0.63\ldots$, which is somewhat below D_{red} for the percolation cluster.

Many more dimensions arise when one discusses transport phenomena
on percolation clusters. In fact, one again observes multifractal behavior.
For a recent review see Aharony (1987).

7.8   Invasion Percolation

Invasion percolation is a *dynamic* percolation process introduced by Wilkinson and Willemsen (1983), motivated by the study of the flow of two immiscible fluids in porous media (de Gennes and Guyon, 1978; Chandler et
al., 1983). Consider the case in which oil is displaced by water in a porous
medium. When the water is injected very slowly then the process takes
place at very low capillary numbers Ca, as discussed in chapter 4. This
implies that the capillary forces completely dominate the viscous forces,
and therefore the dynamics of the process is determined on the pore level.
In the limit of vanishing capillary numbers one may neglect any pressure
drops both in the *invading fluid* (water) and in the *defending fluid* (oil).

However, there is a pressure difference between the two fluids (the capillary
pressure) given by

$$\left(p_{\text{invader}} - p_{\text{defender}}\right) = \frac{2\,\sigma\cos\theta}{r}\ ; \tag{7.38}$$

see also equation (4.4). Here σ is the interfacial tension between the two
fluids, θ is the contact angle between the interface and the pore wall, and
r is the radius of the pore at the position of the interface.

One often finds that water is the 'wetting' fluid and oil the 'nonwetting' fluid, i.e., the contact angle $\theta > 90°$, and the water will spon-
taneously invade the oil-filled porous medium unless the pressure in the
water is kept below that of the defending oil. The important point to note
is that the pressure difference depends on the local radius of the pore or
pore neck where the interface lies. In a porous medium one must have
variations in r (and possibly in the contact angle) and the interface must
adjust to positions so that equation (7.38) is satisfied everywhere. The capillary forces are strongest at the narrowest places in the medium. Thus if
all the throats are smaller than all the pores, the water–oil interface moves
quickly through the throats, but gets stuck entering the larger pores. It is
consistent with both a simple theoretical model and experimental observations to represent this motion as a series of discrete jumps in which at each
time step the water displaces oil from the smallest available pore.

Wilkinson and Willemsen (1983) proposed to simulate this process in
an idealized medium where the network of pores may be viewed as a regular
lattice in which the sites and bonds of the lattice represent the pores and
the throats respectively. Randomness of the medium is incorporated by
assigning random numbers to the sites and bonds to represent the sizes of
these pores and throats. Simulation of the process in a given realization of
the lattice thus consists of following the motion of the water–oil interface
as it advances through the smallest available pore, marking the pores filled
with the invading fluid.

This model also applies in the case where a nonwetting fluid, say air,
displaces a wetting fluid. In this case the pressure in the invading fluid is
above that of the defending fluid and the interface advances quickly through
the large pores and gets stuck in the narrow throats connecting the pores.
An illustration of the types of structures observed in this case is shown in
figure 7.16.

It is apparent in figure 7.16 that the invading fluid *traps* regions of the
defending fluid. As the invader advances it is possible for it to completely
surround regions of the defending fluid, i.e., completely disconnect finite
clusters of the defending fluid from the exit sites of the sample. This is
one origin of the phenomenon of 'residual oil,' a great economic problem

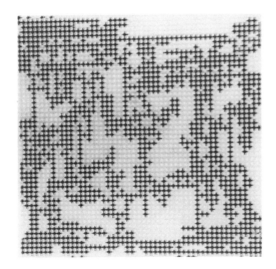

FIGURE 7.16: Invasion percolation of water (black) displacing air in a model consisting of a regular array of cylinders 2 mm in diameter and 0.7 mm high separating parallel plates. The water does *not* wet the model. The water enters at the upper left corner and exits at the lower right corner. The displacement is at $Ca \simeq 10^{-5}$ (Feder et al., 1986).

in the oil industry. Since oil is incompressible, Wilkinson and Willemsen introduced the new rule that water cannot invade trapped regions of oil.

The algorithms describing invasion percolation are now simple to describe:

- Assign random numbers r in the range $[0,1]$ to each site of an $L \times L$ lattice.

- Select sites of injection for the invading fluid and sites of extraction for the defending fluid.

- Identify the *growth sites* as the sites which belong to the defending fluid *and* are neighbors to the invading fluid.

- Advance the invading fluid to the growth site that has the lowest random number r.

- TRAPPING: Growth sites in regions completely surrounded by the invading fluid are not active and are eliminated from the list of growth sites.

- End the invasion process when the invading fluid reaches an exit site.

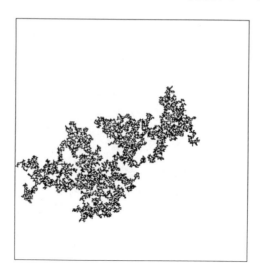

FIGURE 7.17: An invasion percolation (without trapping) cluster grown from the central site on a 300 × 300 lattice until it reaches one of the edges of the lattice. The cluster contains 7656 sites.

This model advances the invading fluid to new sites one by one, always selecting the possible growth site with the lowest random number associated with it. This is an algorithm that lets the invading cluster grow in a manner subject to *local* properties. The rule that a trapped region cannot be invaded introduces a nonlocal aspect into the model. The question whether or not a region is trapped cannot be answered locally, and involves a *global* search of the system.

It is interesting to compare the invasion percolation process without trapping to the ordinary percolation process described in previous sections. In the ordinary percolation process one may grow percolation clusters as follows: The sites on an $L \times L$ lattice are assigned random numbers r in the range $[0,1]$, and one places a seed on the lattice. Then for a given choice of the occupation probability p, $0 \leq p \leq 1$, the cluster grows by occupying all available sites with random numbers $r \leq p$. The growth of the percolation cluster stops when no more such numbers are found on the boundary (perimeter) of the cluster. Of course, most of the sites chosen will generate clusters of a finite size similar to the one shown in figure 7.10. Only if the seed site happens to lie on the incipient percolating cluster at p_c, or on the percolating cluster for $p > p_c$, will the percolation cluster grow to a size that spans the lattice. By contrast, in invasion percolation the cluster grows by always selecting the smallest random number, no matter how large. However, once a large number r_0 has been chosen, it is not

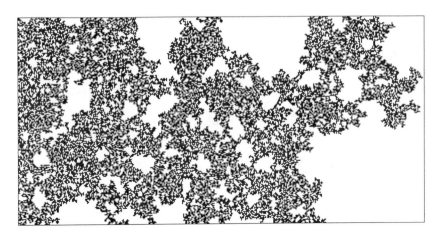

FIGURE 7.18: Invasion percolation (without trapping) on a 100×200 quadratic lattice. The invader (black) enters from sites on the left-hand edge and the defender escapes through the right-hand edge. At 'break-through' the invader just reaches the right-hand edge.

necessarily true that subsequently every number $r \geq r_0$ will be chosen — smaller numbers will in general become available at the interface, and will thus be chosen. The cluster shown in figure 7.17 is grown by this process until it reaches the edge of the system.

The invasion percolation cluster in figure 7.17 is quite similar to the percolation cluster shown in figure 7.10. Wilkinson and Willemsen (1983) simulated the invasion process on lattices of size $L \times 2L$, injecting the invader on the left-hand edge and stopping the simulation at the *break-through* point, when the invader reached the right-hand edge. We illustrate this geometry in figure 7.18.

Naturally, in the finite geometry the invader will gradually fill the entire lattice if the invasion process is continued. Wilkinson and Willemsen found that the number of sites $M(L)$ in the central $L \times L$ portion of the lattice at breakthrough increases with the size of the lattice as follows:

$$M(L) = A L^{D_{\mathrm{inv}}} , \quad \text{with} \quad D_{\mathrm{inv}} \simeq 1.89 . \tag{7.39}$$

This equation is analogous to equation (7.5), and the fractal dimension of invasion percolation without trapping, D_{inv}, is found to equal the fractal dimension of the incipient percolation cluster at p_c. There is now considerable evidence that invasion percolation in fact is in the same universality class as ordinary percolation (Dias and Wilkinson, 1986).

Trapping changes the invasion percolation quite drastically in two dimensions. In figure 7.19 we show an invasion cluster grown by the process

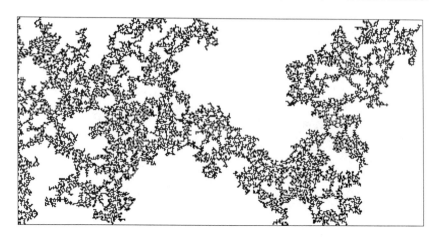

FIGURE 7.19: Invasion percolation with trapping on a 100×200 quadratic lattice. The invader (black) enters from sites on the left-hand edge and the defender escapes through the right-hand edge. At 'break-through' the invader just reaches the right-hand edge.

that includes trapping.   Comparing the invasion percolation cluster in figure 7.18 to the invasion percolation cluster with trapping shown in figure 7.19 one sees that the trapping rule leads to clusters with much larger holes in them. This is reflected in how the number of sites, $M(L)$, that belong to the central part of an $L \times 2L$ lattice, with injection from one side, scales with the size of the lattice,

$$M(L) = A L^{D_{\text{trap}}} , \quad \text{with} \quad D_{\text{trap}} \simeq 1.82 . \tag{7.40}$$

This result was obtained by Wilkinson and Willemsen (1983). Lenormand and Zarcone (1985a) investigated *experimentally* the invasion of air at a very slow rate into a two-dimensional network of 250,000 ducts of random widths on a square grid that was filled by glycerol (see figure 7.20). Note that they continued the experiment after breakthrough. By placing a semipermeable membrane at the right-hand edge of the sample cell they could prevent the invading air from escaping, and the invasion process was terminated when all the remaining defending fluid was trapped. They found that the number of ducts filled by the invading fluid counted inside boxes of side L followed equation (7.40) with $1.80 < D_{\text{trap}} < 1.83$, consistent with the numerical simulations.

We have experimentally generated invasion percolation clusters with the trapping rule by injecting air at the center of a two-dimensional circular model of a porous medium consisting of a layer of glass spheres placed at random and sandwiched between two plates. The resulting cluster, shown

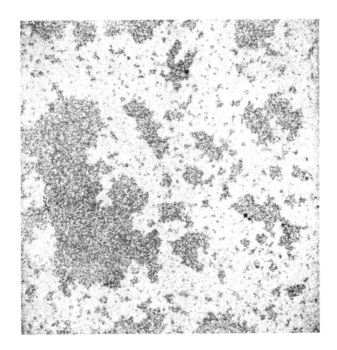

FIGURE 7.20: Displacement of the wetting fluid (black) by the nonwetting fluid (white) injected on the left-hand side of the network. On the right-hand side, a semipermeable membrane prevents the nonwetting fluid from flowing outside (Lenormand and Zarcone, 1985a).

in figure 7.21, was found to have a fractal dimension of $D = 1.84 \pm 0.04$, also consistent with the expected result.

We conclude that the invasion percolation process with trapping generates fractal structures, which have a fractal dimension that is lower than the fractal dimension both of invasion percolation without trapping and of ordinary percolation. The experimental results for immiscible fluid displacement in two-dimensional porous media at very low capillary numbers are consistent with the process of invasion percolation with trapping introduced by Wilkinson and Willemsen (1983).

However, in three dimensions the situation is altogether different. Consider ordinary percolation in the simple cubic (s.c.) lattice where each lattice site has six neighbors. The percolation threshold for this geometry is $p_c(\text{s.c.}) \simeq 0.3117$, and the fractal dimension of the incipient percolation cluster at p_c is $D \simeq 2.5$. The important point to note is that in this case there is a range $p_c \leq p \leq (1 - p_c)$ of occupation probabilities for which *both* the occupied sites *and* the empty sites percolate and form spanning

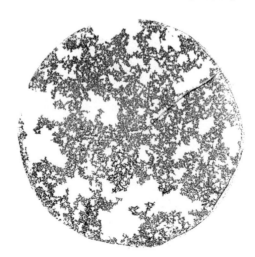

FIGURE 7.21: Air (black) displacing glycerol (white) at very low capillary numbers Ca $\simeq 10^{-5}$, in a two-dimensional model consisting of a layer of randomly packed glass spheres 1 mm in diameter. The fractal dimension of the cluster of invading of air is $D \simeq 1.84$ (Måløy et al., 1987b).

clusters. By contrast in two-dimensional percolation on the square lattice one finds that either the occupied sites or the empty sites percolate and there is *no* range for which both percolate. Interestingly, percolation on the triangular lattice is a borderline case since $p_c = 0.5$ is the simultaneous threshold both for the occupied sites and for the empty sites.

The qualitative difference between two and three dimensions extends to invasion percolation. Wilkinson and Willemsen (1983) find $D \simeq 2.52$ for the invading cluster at breakthrough, both with and without the trapping rule. The existence of a range of occupation probabilities for which both the defending fluid and the invading fluid percolate makes trapping much less effective. Most of the sites are still filled by the defending fluid when the invading fluid percolates the three-dimensional sample from one face to the other. When the invasion process is continued one finds that the defending fluid is trapped at $p = (1-p_c)$. Wilkinson and Willemsen found that at this point the number of sites in the invading cluster increased as L^2, so that it in fact represented a *finite fraction* $\simeq 0.66$ of the sites of the simple cubic lattice of size $L \times L \times L$. The invading cluster is therefore *not* fractal at the limit where the defending fluid is trapped. Dias and Wilkinson (1986) have discussed a related model, *percolation with trapping*, which includes the trapping rule but ignores the invasion part of the problem. They analyze the size distribution of the trapped regions and they give strong evidence for the conclusion that the critical behavior of invasion percolation with trapping belongs to the same universality class as ordinary percolation in three dimensions.

Experimentally it is difficult to realize the three-dimensional invasion percolation process. Clément et al. (1985) injected nonwetting Woods metal into consolidated crushed glass very slowly from the bottom and analyzed both vertical and horizontal cuts of the cylindrical sample. They concluded that the horizontal cuts, perpendicular to the flow direction, show that the Woods metal had invaded the porous medium in a self-similar manner, and the fractal dimension of the distribution of the metal in the cut was found to be $\simeq 1.65$. This is somewhat above the fractal dimension; '1.50, expected for a cross-section of the incipient percolation cluster. However, gravity effects cannot be neglected and clearly influence the results and the fractal dimension of the horizontal cuts depends on the level at which they are taken.

7.9   The Fractal Diffusion Front

We noted in the introduction to this chapter that diffusion processes can spread indefinitely and the dynamics of diffusion lies in the randomness of the particle motion. By contrast in percolation processes the randomness is associated with the *medium*, and there exists a critical threshold below which percolation processes are limited to finite regions or clusters. In a notable paper Sapoval et al. (1985) showed that the *diffusion front* resulting in diffusion from a source has a fractal structure that is related to the so-called *hull* of percolation clusters. The descriptive term hull was first introduced by Mandelbrot (1982) and discussed in detail by Voss (1984).

Consider the diffusion of particles from a line source on a quadratic lattice, shown in figure 7.22. The particles come from a source at the left-hand edge of the illustration. Any one of the particles attempts to jump to one of its four neighboring sites, a distance a away, every τ seconds. The relation between random walks and diffusion in one dimension is discussed in some detail in chapter 9. The diffusion constant $\mathcal{D}$ is given by the Einstein relation (9.3) which in the notation used here is

$$\mathcal{D} = \frac{1}{2\tau} a^2 \; . \tag{7.41}$$

The displacements of the diffusing particle in the x-direction, perpendicular to the source, and in the y-direction, parallel to the line source, are statistically independent. The mean square displacement, during a time interval t, of a particle starting at x_0, y_0 is given by equation (9.11) or for

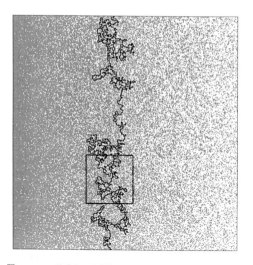
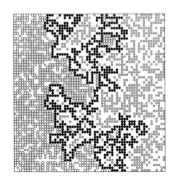

FIGURE 7.22: Diffusion of 'particles' (filled circles) from a source at the left-hand edge of a lattice consisting of sites (empty squares). The particles (filled squares) that are both connected to the source and neighbors to sites connected to the right-hand edge constitute the *hull* of the diffusing particles. The right-hand figure is a blow-up of the region marked by a square in the left-hand figure. Here the hull is shown as filled circles, sites connected to the source as large circles and the remaining occupied sites as small circles. A color version of the left-hand part of this figure is included in the insert that follows the Contents.

the two-dimensional case

$$\langle [x(t) - x_0]^2 \rangle = \langle [y(t) - y_0]^2 \rangle = 2\mathcal{D}t \tag{7.42}$$

The diffusion distance ℓ is defined as the root mean square displacement of the diffusing particle from its starting point and is given by

$$\ell^2 = \langle [x(t) - x_0]^2 \rangle + \langle [y(t) - y_0]^2 \rangle = 4\mathcal{D}t = 2a^2t/\tau . \tag{7.43}$$

It is well known that the probability of finding a particle at position x from the line source, at $x = 0$, in a lattice of width L and infinite in the x-direction is given by

$$p(x) = 1 - \frac{2}{\sqrt{\pi}} \int_0^{x/\ell} du \exp(-u^2) . \tag{7.44}$$

This probability decreases gradually from 1 at the source and vanishes rapidly for $x > \ell$.

The illustration in figure 7.22 shows that when the diffusion process is considered at a given instant in time one finds a lattice with sites occupied by particles just as in the percolation process discussed in previous sections. However, in this case the probability of a site being occupied is dependent on the distance x from the source and is given by equation (7.44). Near the source one has $p(x) \simeq 1$, which is above the percolation threshold of the quadratic lattice, and the sites occupied by particles percolate. Further away from the source the probability of having sites occupied by particles falls below p_c and these sites form only isolated clusters.

In an electrical analog one considers particles on neighboring sites to be in electrical contact. As an example imagine gold atoms diffusing from a source on an insulating substrate. The border between the sites that are (electrically) connected to the source and the insulating nonoccupied sites is called the *hull* of that region. Some care has to be taken in the definition of the hull. Two sites occupied by particles are connected to other empty sites if they are neighbors in the x- or y-directions. We also consider the connectivity of the empty sites. However, one must (in order to obtain a well-defined border between the filled sites and the empty sites) consider an empty site to be connected if at least one of its eight neighboring sites are also empty. That is, in defining the connectivity of the empty sites we consider not only the neighboring sites in the x- and y-direction but also the four diagonally placed neighboring sites. The hull consists of all the sites that are connected to the source *and* are neighbors to empty sites connected to the insulating far end of the sample. We consider, in a pictorial language, the sites connected to the source to represent 'land,' and the connected empty sites to represent the 'ocean.' One walks along the hull if one makes sure that there is salt water just next to one's position at all times. Note that this excludes the shores of the lakes of disconnected empty sites on land (they are not salty), and it also excludes the shores of islands off the 'hull' and in the ocean since they are not connected to the source.

In figure 7.22 we show the result of a simulation on a 300×300 lattice, after a time of $t = 9 \cdot 10^4 \tau$, and with a diffusion length of $\ell \simeq 300a$ (for a color version of this figure see the insert that follows the Contents). The position of the hull moves to increasing x as time, and therefore $\ell = a\sqrt{2t/\tau}$, increases. Also the width of the region covered by the hull increases with time and ℓ. To be more precise, let $p_h(x)dx$ be the probability that a site in the range from x to $x + dx$ belongs to the hull. Then the position of the hull, x_h, and the width σ_h are given by

$$x_h = \int_0^\infty dx\, x\, p_h(x)\ ,$$
$$\sigma_h^2 = \int_0^\infty dx\, (x - x_h)^2\, p_h(x)\ . \tag{7.45}$$

Sapoval et al. made the important observation that the position of the hull, x_h, is given by

$$p(x_h) = p_c \qquad (7.46)$$

This implies that by finding the probability p of having a site occupied a distance x_h from the source one obtains a very accurate estimate of the percolation threshold. For the two-dimensional quadratic lattice Rosso et al. (1985) obtained the value $p_c = 0.592802 \pm 10^{-5}$, and Ziff (1986) found by an analysis of percolation cluster hulls that $p_c = 0.59275 \pm 3 \cdot 10^{-5}$. For the triangular lattice Sapoval et al. (1985) found $p_c = 0.5011 \pm 0.0003$, consistent with the exact result $p_c = 1/2$.

When the hull is analyzed over distances smaller than the width σ_h of the hull Sapoval et al. found that the number of sites, $M(R)$, that belong to the hull inside a radius R scales as

$$M(R) \sim R^{D_h} , \quad \text{with} \quad D_h = 7/4 . \qquad (7.47)$$

The hull is a fractal object. In fact they found $D_h = 1.76 \pm 0.02$, by fitting the results of simulations. However, further analysis of these simulations (to be discussed below) led them to conjecture that $D_h = 1.75$, exactly. On length scales less than the width of the hull, it appears to be a self-similar fractal with a fractal dimension $\simeq 1.75$. This result is consistent with the determination of the fractal dimension of the hulls of percolation clusters by Voss (1984). He estimated $D_h = 1.74 \pm 0.02$ from clusters of varying size, and $D_h = 1.76 \pm 0.01$ from the two-point correlation of the sites on the hull. Ziff (1986) concluded by studying hulls of percolation clusters that the fractal dimension of the hull is 1.751 ± 0.002, consistent with the conjecture by Sapoval et al.

The self-similarity of the hull is shown very nicely in figure 7.23. It is clear, however, that the hull in the geometry used here is in fact a self-affine fractal. The distinction between self-similar and self-affine is discussed in chapter 10. The point is that if one considers the diffusion on a strip of width $2L$ instead of L, for a given time t, then one finds that the number of sites that belong to the hull also doubles. From this point of view the diffusion front is a one-dimensional object. Sapoval et al. therefore studied the scaling properties of the number of sites in the hull, $M_h(L, \ell)$, and of the width of the hull, $\sigma(L, \ell)$, by fitting the results of simulations using the relations

$$
\begin{aligned}
M_h(L, \ell) &= A\, L\, \ell^{\alpha} , \quad \text{with} \quad A = 0.96 \quad \text{and} \quad \alpha = 0.425 \pm 0.005 , \\
\sigma_h(L, \ell) &= B\, \ell^{\hat{\alpha}} , \quad \text{with} \quad B = 0.46 \quad \text{and} \quad \hat{\alpha} = 0.57 \pm 0.01 .
\end{aligned}
\qquad (7.48)
$$

The results of the simulations and the fits of equation (7.48) are shown in figure 7.24.

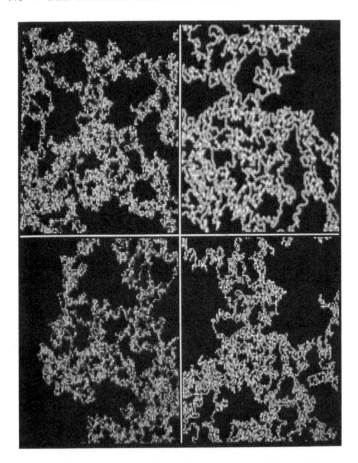

FIGURE 7.23: Self-similarity of the diffusion front. The hull is shown as a series of magnifications of a central part. The magnification is a factor of two each time. The self-similarity is apparent in the fact that it calls for a rather close inspection of the four figures if one wants to order the figures according to magnification (Sapoval et al., 1986).

Sapoval et al. gave the following interesting argument relating the exponent $\hat{\alpha}$ to the exponent ν for the correlation length: When x does not equal position x_h of the front, then the concentration of sites occupied by particles deviates from p_c. There are lakes and islands in these regions. The characteristic size of these objects is the correlation length ξ, which depends on position in this case, and is given by a modified form of equation (7.23):

$$\xi = \xi_0 |p(x) - p_c|^{-\nu} . \tag{7.49}$$

If the distance to x_h of a cluster (lake or island) of characteristic size $\xi(x)$

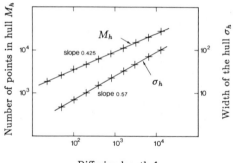

FIGURE 7.24: Variation of the number of points in the hull, M_h, and of the width σ_h, as functions of the diffusion length ℓ (Sapoval et al., 1985).

is equal to ξ this cluster has a finite probability of touching the frontier and thus of being within the frontier. In other words

$$\sigma_h = K\,\xi(x_h \pm \sigma_h)\;. \qquad (7.50)$$

Here K is a constant of order unity. Here equation (7.49) may be used, and with a Taylor expansion of $p(x)$ given by equation (7.44), one finds that

$$\sigma_h = K\xi_0 \left[\sigma_h\left(\frac{\partial p}{\partial x}\right)_{x=x_h}\right]^{-\nu} \qquad (7.51)$$

From equation (7.44), one finds that $(\partial p/\partial x)_{x=x_h} \propto 1/\ell$, and obtains

$$\sigma_h = K'\xi_0 \cdot (\sigma_h/\ell)^{-\nu}\;, \qquad (7.52)$$

It follows that

$$\sigma_h \sim \ell^{\hat{\alpha}} \sim \ell^{\frac{\nu}{1+\nu}}\;. \qquad (7.53)$$

The result is that the exponent $\hat{\alpha}$ that controls how the width of the hull increases with the diffusion length ℓ is given in terms of the exponent ν that controls the divergence of the correlation length by

$$\hat{\alpha} = \frac{\nu}{1+\nu}\;. \qquad (7.54)$$

If we use the exact value $\nu = 4/3$ we predict $\hat{\alpha} = 4/7 = 0.5714$, while the simulations gave $\hat{\alpha} = 0.57 \pm 0.01$.

We have already noted that the hull is a self-similar fractal up to length scales equal to the width of the hull. Therefore we expect the number of sites that belong to the hull in a box of size $\sigma_h \times \sigma_h$ to be given by

$$M_h(L,\ell)\frac{\sigma_h}{L} \sim \sigma_h^{D_h} = \ell^{\alpha}\ell^{\frac{\nu}{1+\nu}}\;. \qquad (7.55)$$

Here the last relation follows from equations (7.48) and (7.54), and one is led to the conclusion

$$\alpha = \frac{\nu}{1+\nu}(D_h - 1) = 3/7 = 0.429\ldots , \qquad (7.56)$$

where we have used $D_h = 7/4$ and $\nu = 4/3$. The observed value $\alpha = 0.425 \pm 0.005$ agrees very well with the prediction.

Sapoval et al. (1985) noted that the ratio $M_h(L/\ell)\sigma_h/L\ell$ approaches a constant value of $\simeq 0.441$ as $\ell \to \infty$. This result together with the scaling result in equation (7.55), led them to conjecture that the fractal dimension of the hull can be written

$$D_h = \frac{1+\nu}{\nu} = 7/4 = 1.75 . \qquad (7.57)$$

This result is consistent with the results of large-scale simulations by Ziff (1986). Very recently Saleur and Duplantier (1987) proved this conjecture to be correct!

In many cases one is interested not in the hull of a cluster but in the *external perimeter* (Grossman and Aharony, 1986). The external perimeter includes the sites available to a finite-size particle (coming from the outside) that is touching the occupied sites on the cluster. This is what is required if we consider the adsorption of finite-size particles on a fractal surface. The external perimeter differs from the hull in that many fjords cannot be reached by the test particle. Grossman and Aharony (1986) have shown that the fractal dimension of the external perimeter is $D_e = 1.37 \pm 0.03$. Heuristic arguments show that $D_e = 4/3$, exactly (Aharony, 1986; Saleur and Duplantier, 1987; Grossman and Aharony, 1987). Grossman and Aharony (1987) extended the definition of the external perimeter to the *accessible perimeter*, defined as all the perimeter sites that are connected to infinity by a channel (of empty sites) which has a minimal width larger than r (the diameter of the test particle). They found by numeric simulations that the fractal dimension of the accessible perimeter is also $D_e = 4/3$ independent of r when the test-particle diameter is larger than some lattice-dependent threshold.

Shaw (1987) studied the displacement front formed by the evaporation of water from a quasi two-dimensional porous medium. He found that the front is stable and locally has a structure characteristic of invasion percolation (see figure 7.25a), and he determined a fractal dimension $D = 1.89 \pm 0.03$ for the front. In a lower magnification (see figure 7.25b) the front is quite similar in structure to the hull shown in figure 7.22. An analysis of the leading edge, i.e., the perimeter of the front, shows that the perimeter has a fractal dimension $D = 1.38 \pm 0.02$, which is consistent with the expected value for D_e.

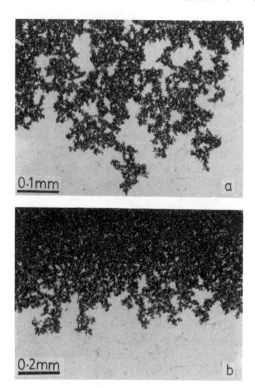

FIGURE 7.25: The drying front in a thin layer of silica spheres. Drained porosity appears dark in the transmitted light image. (a) Portion of the drying front. (b) In the lower-magnification image it is evident that the front is stable (Shaw, 1987).

The discovery by Sapoval, Rosso and Gouyet that the diffusion front has a fractal structure is notable. Diffusion has been studied for a long time and is completely 'understood' in terms of the diffusion equation, which leads to time-dependent diffusion fronts such as that described by equation (7.44). Nevertheless the diffusion front has an internal structure that is fractal. It should also be remembered that the fractal structure extends over distances comparable to the diffusion width $\ell = \sqrt{4\mathcal{D}t}$, which diverges with time, and the fractal structure may very well extend over macroscopic distances even in the case where the diffusion is on the atomic scale.

Rosso et al. (1986) have extended this discussion to three-dimensional diffusion on the simple cubic lattice. We have already, in the context of invasion percolation, studied the fact that the connectivity properties in two- and three-dimensional systems are qualitatively different (see figure 7.26).

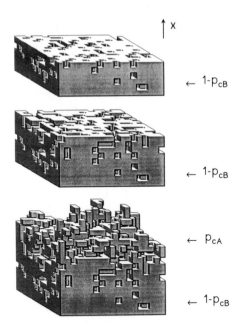

FIGURE 7.26: Picture of a $19 \times 19 \times 19$-site system. The particles are shown as cubes, and the probability of having a site occupied decreases from 1 at the bottom of the sample to 0 at the top. Only sites occupied by particles connected to the bottom are shown. (a) The top figure shows the bottom six layers in the high-density part of the sample. (b) The top layer in the middle figure is at a concentration corresponding to the percolation threshold of the quadratic lattice. (c) The bottom figure shows all the sites connected to the bottom plane. The arrows indicate the planes where the occupation probability is given by $1 - p_{cB}$ and by p_{cA} (Rosso et al., 1986).

The diffusing particles are considered to be connected if they are nearest neighbors, i.e., they are neighbors in the x-, y- or z-directions. The percolation threshold for site percolation with this connectivity is $p_{cA} \simeq 0.3117$. An empty (insulating) site is considered to be connected to another empty site if it is on one of the 26 neighboring sites in a cube of side $3a$, centered on it. The percolation threshold for the empty sites with this connectivity is $p_{cB} \simeq 0.097$. In figure 7.26 one clearly sees that it is possible to have percolation of both the particles and the holes simultaneously in a range of occupation probabilities $p_{cA} \leq p \leq (1 - p_{cB})$.

The hull of the occupied sites connected to the bottom plane includes the sites that are on the cluster connected to the bottom plane and have as a nearest neighbor an empty site connected to the top plane. Interestingly, for p in the range of simultaneous percolation, one finds that almost all the occupied sites belong to the hull. The *surface* of the diffusion front, which is the hull, is in fact a finite fraction of the sites, and therefore an object of fractal dimension 3. From this point of view it behaves as an ordinary solid. However, any point of that solid can be reached from the outside, i.e., it belongs to the surface. In that sense, this system is an ideally porous material.

Chapter 8

Fractal Records in Time

Many observations of nature consist of records in time or a series of observations. For example, extensive records exist for temperature. These records clearly exhibit yearly variations. Long records of temperature show erratic behavior on both a short- and a long-term time scale. The records in time of such phenomena as temperature, the discharge of rivers, rainfall and thickness of tree rings can be analyzed in terms of *Hurst's rescaled range analysis*. The records are characterized by an exponent H — the Hurst exponent. The trace of the record is a curve with a fractal dimension $D = 2 - H$, under conditions we will discuss in more detail in chapter 10.

In this chapter we present and discuss Hurst's analysis of records in time. The related problem of fractal Brownian motion is discussed in the next chapter. We use the rescaled range analysis in a discussion of ocean wave-height statistics, after a discussion of the relation between self-similar and self-affine curves.

8.1 Hurst's Empirical Law and Rescaled Range Analysis

Hurst spent a lifetime studying the Nile and the problems related to water storage. He invented a new statistical method — the *rescaled range analysis* (R/S analysis) — which he described in detail in an interesting book, *Long-Term Storage: An Experimental Study* (Hurst et al., 1965). As an introduction to this method let us consider Lake Albert, an example discussed by Hurst. In figure 8.1, we have plotted the measured annual discharge as a function of time.

The problem is to determine the design of an ideal reservoir based upon the given record of observed discharges from the lake. An ideal reservoir

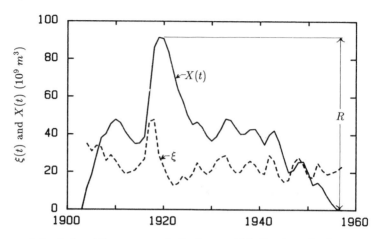

FIGURE 8.1: Lake Albert annual discharge $\xi(t)$ (broken line), and accumulated departures from the mean discharge $X(t)$ (full line). The range is indicated by R (after Hurst et al., 1965).

never overflows or empties. In any given year, t, such a reservoir will accept the influx $\xi(t)$ from the lake, and a regulated volume per year (discharge), $\langle \xi \rangle_\tau$, will be released from the reservoir. What storage would have been required for the reservoir to release a volume each year equal to the mean influx for the period under discussion? The average influx over the period of τ years is

$$\langle \xi \rangle_\tau = \frac{1}{\tau} \sum_{t=1}^{\tau} \xi(t) \, . \tag{8.1}$$

This average should equal the volume released per year from the reservoir. Let $X(t)$ be the *accumulated* departure of the influx $\xi(t)$ from the mean $\langle \xi \rangle_\tau$

$$X(t, \tau) = \sum_{u=1}^{t} \{ \xi(u) - \langle \xi \rangle_\tau \} \, . \tag{8.2}$$

The resulting curve for Lake Albert is shown in figure 8.1. The difference between the maximum and the minimum accumulated influx X is the *range* R. The range is the storage capacity required to maintain the mean discharge throughout the period. For a sufficiently large reservoir that never overflows and never empties, R represents the difference between the maximum and minimum amounts of water contained in the reservoir. The explicit expression for R is

$$R(\tau) = \max_{1 \leq t \leq \tau} X(t, \tau) - \min_{1 \leq t \leq \tau} X(t, \tau) \, , \tag{8.3}$$

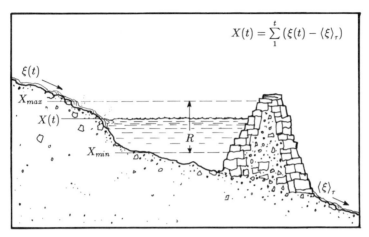

$$X(t) = \sum_{1}^{t} (\xi(t) - \langle \xi \rangle_\tau)$$

FIGURE 8.2: Sketch of a reservoir with an influx of $\xi(t)$, and an average discharge $\langle \xi \rangle_\tau$. The accumulated difference between influx and regulated outflow is $X(t)$. The range, R, is the difference between the maximum and minimum contents of the reservoir.

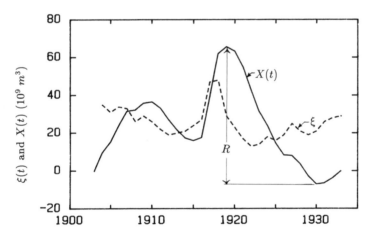

FIGURE 8.3: Lake Albert accumulated departures from the mean discharge $X(t)$ for the first 30 years. The range is indicated by R (after Hurst et al., 1965).

where t is a discrete integer-valued time and τ is the time-span considered. These quantities are illustrated in figure 8.2.

Clearly, the range depends on the time period τ considered and we expect the range R to increase with increasing τ. For the Lake Albert data in figure 8.1 for the period 1904 to 1957 one finds $R(53) = 91 \cdot 10^9 \, \text{m}^3$,

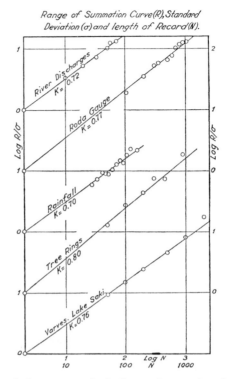

FIGURE 8.4: Rescaled range analysis for various natural phenomena. The lag is $\tau = N$, the number of years (Hurst et al., 1965).

whereas for the first 30 years, illustrated in figure 8.3, one finds a range of only $R(30) = 73 \cdot 10^9\,\text{m}^3$.

Hurst investigated many natural phenomena, such as river discharges, mud sediments and tree rings. Hurst used the dimensionless ratio R/S, where S is the standard deviation, i.e., the square root of the variance. The use of this dimensionless ratio permits the comparison of observed ranges of various phenomena. The standard deviation is estimated from the observations by

$$S = \left(\frac{1}{\tau} \sum_{t=1}^{\tau} \{\xi(t) - \langle \xi \rangle_\tau\}^2 \right)^{\frac{1}{2}} . \tag{8.4}$$

Hurst found that the observed rescaled range, R/S, for many records in time is very well described by the following empirical relation:

$$R/S = (\tau/2)^H . \tag{8.5}$$

Properties of K from Natural Phenomena

Phenomenon	Range of N Years	Number Pheno-mena	Sets	Mean	K Std. devn.	Range	Coeff. of auto-correl-ation
River discharges	10–100	39	94	0·72	0·091	0·50–0·94	
Roda Gauge	80–1,080	1	66	0·77	0·055	0·58–0·86	0·025±0·26
River and lake levels	44–176	4	13	0·71	0·082	0·59–0·85	$n=15$
Rainfall	24–211	39	173	0·70	0·088	0·46–0·91	0·07±0·08* $n=65$
Varves							
Lake Saki	50–2,000	1	114	0·69	0·064	0·56–0·87	−0·07±0·11
Moen and							$n=39$
Tamiskaming	50–1,200	2	90	0·77	0·094	0·50–0·95	
Corintos and							
Haileybury	50–650	2	54	0·77	0·098	0·51–0·91	
Temperatures	29–60	18	120	0·68	0·087	0·46–0·92	
Pressures	29–96	8	28	0·63	0·070	0·51–0·76	
Sunspot numbers	38–190	1	15	0·75	0·056	0·65–0·85	
Tree-rings and spruce index	50–900	5	105	0·79	0·076	0·56–0·94	
Totals and means of sections							
Water statistics		83	346	0·72	0·08	0·46–0·94	
Varves		5	258	0·74	0·09	0·50–0·95	
Meteorology and trees		32	268	0·72	0·08	0·46–0·94	
Grand totals and means	10–2,000	120	872	0·726	0·082	0·46–0·95	

* Includes also river discharges.

TABLE 8.1: Table of $H(= K)$ for various natural phenomena (Hurst et al., 1965).

The *Hurst exponent* H (called K by Hurst[1]), is more or less symmetrically distributed about a mean of 0.73, with a standard deviation of about 0.09.

A figure from Hurst's book illustrates the quality of the fit of the *Hurst empirical law* equation (8.5) to observations — see figure 8.4. Statistical results collected by Hurst are found in table 8.1, which clearly shows that for many natural phenomena we have $H > 1/2$.   Hurst's observation is remarkable considering the fact that in the *absence* of long-run statistical dependence R/S should become asymptotically proportional to $\tau^{\frac{1}{2}}$ for records generated by statistically independent processes with finite vari-

---

[1]Mandelbrot used H for the Hurst exponent. This is a fortunate choice since H is often directly related to the Lipschitz-Hölder exponent.

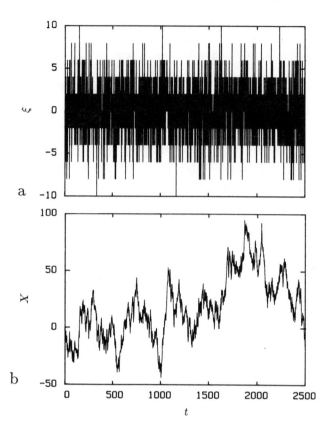

FIGURE 8.5: (a) A sequence of independent random numbers $\xi(t)$ generated by 'flipping' 10 'coins' 2500 times. (b) The accumulated deviation from the mean (zero) $X(t) = \sum_{u=1}^{t} \xi(u)$.

ances and is given by $R/S = 1.2533\,\tau^{1/2}$

$$R/S = (\pi\tau/2)^{\frac{1}{2}}$$ independent random process with finite variances,

(8.6)

as shown by Hurst (1951) and Feller (1951).

8.2   Simulations of Random Records

Hurst — apparently a practical (and skeptical) man — tested equation (8.6) by 'Monte Carlo' simulation for a process of independent random variables obtained by tossing n coins a total of τ times and taking the random variable to be $\xi = $ (number of heads) − (number of tails). The probability

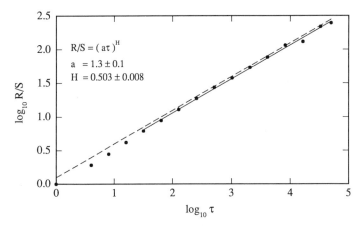

FIGURE 8.6: R/S as a function of lag τ (filled circles), for a random variable $\xi(t)$ given as the difference between the number of 'heads' and 'tails' in tosses of 10 coins repeated $T = 50,000$ times. The Gaussian asymptotic behavior $R/S = \sqrt{\pi\tau/2}$, is shown as the broken line. The line represents a fit of $R/S = (a\tau)^H$ to the observed R/S. The parameters of the fit are $a = 1.3 \pm 0.1$ and $H = 0.503 \pm 0.008$.

of obtaining k heads by throwing n coins is $(\frac{1}{2})^n(n!/k!(n-k)!)$. If the set is tossed τ times then k, and therefore ξ, are given by the binomial distribution, which approaches the normal or Gaussian distribution for large τ and n. It is straightforward (Hurst, 1951) to show that for this process one finds

$$R = \sqrt{\frac{\pi}{2}n\tau} - 1$$

Since the standard deviation of the number of heads minus the number of tails is twice the standard deviation of k, and is given by $S = \sqrt{n}$, we find that equation (8.6) follows in the limit of large τ.

Hurst made experiments by tossing 10 coins 1000 times — it took him about 35 minutes to toss 10 coins 100 times! We have simulated this process on a computer using a (pseudo-)random number generator to select -1 and 1 with equal probability. We consider the heads to be represented by the 1's. From $n = 10$ selections we evaluate $\xi(t)$ as the sum of the numbers generated. The process is repeated 2500 times — this takes less than a second. The resulting sequence of random independent variables $\xi(t)$ is plotted in figure 8.5. The sequence $\xi(t)$ looks like *noise*. We have drawn lines between the points at $(t, \xi(t))$ and $(t-1, \xi(t-1))$, for $t = 1, \ldots, 2499$, in order to give a readable representation of the data. The accumulated departure from the mean $X(t)$ is also shown in figure 8.5. We have again connected discrete points, $(t, X(t))$, by lines that represent the *record* of the set of points that have been visited. We note that the record $X(t)$ is the position at time t of a particle that walks at random with steps of unit length on a line. This random process is a simplified version of the random walk with a Gaussian distribution of the step length discussed in the next chapter. It may be shown that, on time-scales much larger than the time between steps and on distances much larger than the unit step length, the present random walk with unit steps becomes asymptotically an ordinary Brownian motion.

We have calculated R/S for data of the type shown in figure 8.5 starting with a sequence of $\tau = T = 50,000$ 'flips' of 10 'coins.' The time-span, τ, over which the record in time is analyzed is called the *lag*. We reduce the lag τ by a factor of 2 and obtain two independent values of R/S, from the record in time, one for each half of the data. We proceed to reduce τ by another factor of 2, until we have $\tau < 8$, doubling the number of independent domains in each step. For $\tau = 1$ it follows from the definitions (8.3) and (8.4) that $R/S = 1$.

The results corresponding to the same value of the lag τ are averaged and are plotted in a double logarithmic plot of R/S as a function of lag τ in figure 8.6.

The asymptotic R/S expected for independent random variables given by equation (8.6) is shown as the dashed line in figure 8.6. We find that the simulated results are well described by the asymptotic form for $\tau > 20$, and fall significantly below for $\tau < 20$. A least-squares fit of the observed R/S with the form $(a\tau)^H$ in the range $\tau > 20$ gives the estimates $a = 1.3 \pm 0.1$ and $H = 0.503 \pm 0.008$, consistent with the asymptotic form $(1.57\tau)^{\frac{1}{2}}$. The errors are the standard deviations estimated from the covariance matrix obtained in the fit. The error given thus describes how well the line fits the datapoints. Normally, one finds that the variations between H estimated in different runs is somewhat larger than the quoted errors. For example, if we include all points $\tau \geq 4$ in the fit, then we find $a = 1.04 \pm 0.08$, and $H = 0.516 \pm 0.006$, which overestimates H and underestimates a. As already pointed out by Mandelbrot and Wallis (1969a), Hurst, in using the empirical Hurst law $R/S = (\tau/2)^H$, tends to overestimate H for $H < 0.72$ and underestimate H for $H > 0.72$. It is fair to point out that Hurst was fully aware of this and uses his simple form only because there is not enough data to warrant a more complicated fit of the type we have done — he states in connection with his coin toss experiments: *'Short records with low values of K are therefore not distinguishable from random records.'*

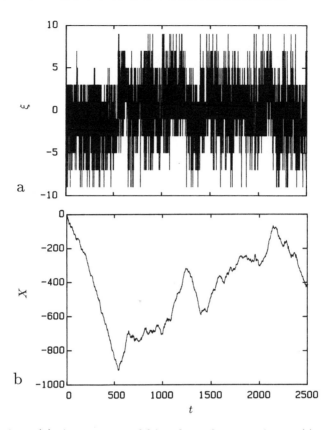

FIGURE 8.7: (a) A sequence of biased random numbers $\xi(t)$ generated by drawing 'cards' from a biased 'hand' generated by the Hurst process, repeated 2500 times. (b) The accumulated deviation from the (zero) mean $X(t) = \sum_{u=1}^{t} \xi(u)$.

8.3   Simulations of Long-Term Dependence

In trying to account for the R/S statistics with $H \sim 0.72$, Hurst made simulations using a 'probability pack of cards.' In this pack cards are numbered $-1, +1, -3, +3, -5, +5, -7, +7$, and the numbers of each kind are proportional to the ordinates of a normal frequency curve. There are 52 cards in all: thirteen 1's, eight 3's and four and one respectively of the others. The approximation of these numbers to the Gaussian normal frequency curve is fairly close. The cards are first well shuffled and then cut, and the number on the exposed card recorded. The cards are reshuffled slightly and cut again, and so on. The numbers recorded may be taken as

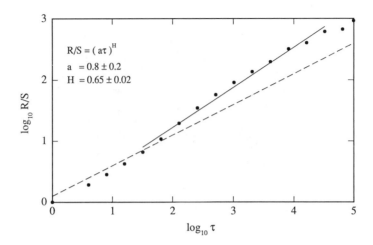

FIGURE 8.8: R/S as a function of the lag τ (filled circles) for a random variable $\xi(t)$ generated by cutting cards from a biased hand by the Hurst procedure, repeated $T = 100,000$ times. The Gaussian asymptotic behavior $R/S = \sqrt{\pi\tau/2}$ is shown as the broken line. The fit (full line) of the form $R/S = (a\tau)^H$ to the observed values of R/S shown was obtained with the parameter values $a = 0.8 \pm 0.2$, and $H = 0.65 \pm 0.02$.

corresponding to observations of a quantity whose frequency distribution conforms to the normal Gaussian curve.

This process is quicker than tossing coins — Hurst produced 100 random numbers in 20 minutes this way. Simulating this process we find results very similar to those shown in figures 8.5 and 8.6, as is to be expected.

Hurst then made an interesting extension of his simulations producing *biased* random records:

> *The pack is shuffled and a card is cut, and after its number has been noted it is replaced in the pack. Two hands are then dealt and if for example the card cut was +3, then the three highest positive cards in one hand are transferred to the other, and from this the three highest negative cards are removed. This hand then has a definite bias. A joker is now placed in it.*

Hurst then uses this biased probability hand of cards to generate a random sequence as before. If the joker is cut, all the cards are reshuffled and a new biased hand dealt.

Hurst made six such experiments, each consisting of 1000 cuts, and determined an exponent $H = 0.714 \pm 0.091$, which is consistent with his observations on the long-term statistics of natural phenomena!

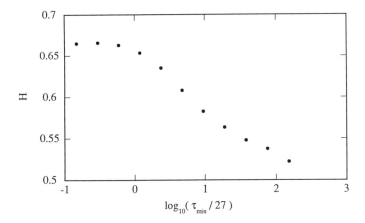

FIGURE 8.9: The apparent Hurst exponent H as a function of the minimum lag $\tau_{\min}$ used in fitting a trend line to the R/S versus τ data in figure 8.9, for a sequence of $100,000$ dependent variables generated by the biased Hurst process.

We have simulated a Hurst biased random walk, and the resulting record of the random variable $\xi(t)$ in figure 8.7a is quite different from the corresponding record for the independent random sequence in figure 8.5a. The accumulated deviation from the mean $X(t) = \sum_{u=1}^{t} \xi(t)$ shown in figure 8.7b shows large excursions with less 'noise.' Thus, for small values of the lag τ, the range R divided by the *sample average* S is smaller than the value found for the independent process in figure 8.5b. However, for τ above 100 the R/S is substantially above the independent values. Fitting the observed set of R/S in the range $20 < \tau < 2500$, we find $a = 0.62 \pm 0.07$ and an exponent $H = 0.71 \pm 0.01$, consistent with Hurst's simulations using a probability pack of cards.

It is clear that the biased Hurst process generates trends that remain for, on the average, $\tau = n$ cuts of a hand containing n cards. For the present case one must on the average cut the hand 27 times before the joker appears. Thus if the hand has a positive bias the trend will be increasing, whereas a negatively biased hand will give a decreasing trend. In the long run we expect the random sequence produced in this way to behave as an independent random process — with the asymptotic behavior given by equation (8.6) as before.

In figure 8.8 the R/S analysis of a biased Hurst process with $100,000$ cuts of the probability pack described above is given. The *trend line* obtained from fitting the data for $\tau > 20$ gives $a = 0.8 \pm 0.2$ and an *apparent* Hurst exponent of $H = 0.65 \pm 0.02$. Clearly the trend line does not fit the

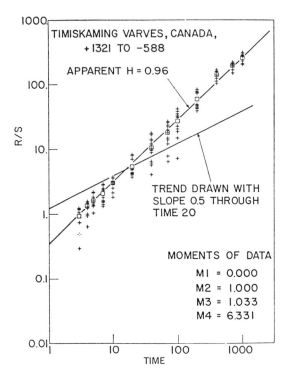

FIGURE 8.10: R/S diagram for the thickness of the varves in Timiskaming, Canada, as a function of the lag τ in years (Mandelbrot and Wallis, 1969a).

results of the simulation too well. In fact if we evaluate the *apparent* Hurst exponent by fitting the results in the range $\tau_{\min} \leq \tau \leq 4096$ we find that the estimated H drops markedly when $\tau_{\min} \sim 27$ — the average residence time of the joker in the biased hand (see figure 8.9).

The approach to the Gaussian asymptote is very slow. In fact the apparent H approaches $H \simeq 0.5$ as $\tau_{\min}$ increases, but the estimated values for H become rather uncertain since several decades of data are necessary for accurate fits. We conclude that the asymptote will not be reached in simulations unless the lag is extended to *very* large samples. We find it, with this background, very difficult to assess the significance of Hurst exponents $H \neq 1/2$ estimated from a limited set of observations.

One of the longest records analyses using the rescaled range method has been done by Mandelbrot and Wallis (1969a), in studying the fossil weather record in the form of thicknesses of mud layers in varves in Timiskaming, Canada. The data stretch over a period of 1809 years, have a very high apparent Hurst exponent $H = 0.96$ and show no evidence of breaking away from this trend line (see figure 8.10).

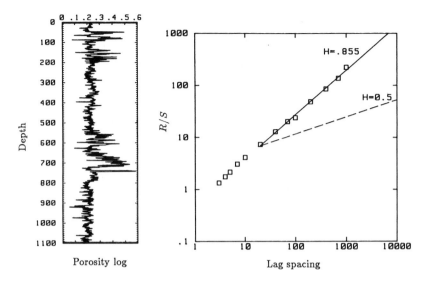

Porosity log   Lag spacing

FIGURE 8.11: Porosity as a function of depth (feet) measured using a well log (left figure). R/S as a function of lag for the porosity data (right figure). The trend line is a fit with $H = 0.855$. The broken line is the slope for a statistically independent process (Hewett, 1986).

Why should natural phenomena show the Hurst statistics? This clearly is an open question. However, the biased Hurst walk with a joker in the pack of probability cards gives a clue. The discharge of a river depends not only on the recent precipitation but also on earlier rainfalls. The flow in a large river system such as the Nile, or the discharge of Lake Albert, must depend on the water content in a large drainage area. The amount of water stored in the drainage area will increase in prolonged periods of higher than average precipitation. The excess amount of water stored will then contribute to the discharge in drier years. If for a prolonged period of time there is less than normal rainfall the general level of water in the drainage basin falls, and in subsequent periods of high precipitation some of the water is absorbed by the drainage area and the discharge remains less than normal. These 'memory' effects are modeled by the joker in Hurst's pack of cards. For river discharges the fractal nature of the drainage area, as discussed in section 12.2, may also contribute to the fractal behavior of river discharges. The fractional Brownian motion model (Mandelbrot and van Ness, 1968) discussed in the next chapter in effect takes memory effects into account.

We have shown that an extraordinarily long series of observations must be used in order to obtain the Gaussian statistics for systems with even a moderate memory effect. Therefore it is not clear to what extent Hurst exponents $H > \frac{1}{2}$ obtained from observations by the R/S method imply persistence — and again more research is needed.

As an example of a recent R/S analysis consider the results reproduced in figure 8.11. The porosity as a function of depth was logged in a well (Hewett, 1986). The figure shows that the porosity fluctuates strongly. The R/S analysis shown in the right-hand part of figure 8.11 clearly indicates a persistent behavior with a Hurst exponent of $H = 0.855$. Hewett also analyzed the variance of increments and found the same value of H.

Chapter 9

Random Walks and Fractals

Randomness is inherent in all natural phenomena. Even the most perfect crystal has many impurities and other defects placed at random. In fact, even if the crystal was perfect with each atom in its proper place, it would be there only on the average since the atoms are in constant thermal motion. Therefore the actual state of even the most perfect system has elements of randomness. There is good evidence that many natural phenomena are best described as fractals. However, if fractals are to be useful in the description of nature we must develop the concepts of *random fractals*.

Because of the extraordinary importance of *Brownian motion*, or the *random walk* process, in physics, chemistry and biology we will start with a discussion of this process as an example of random processes with fractal properties. The simplest version is the one-dimensional random walk, which then may be extended to higher dimensions. We also consider the generalization to *fractional* Brownian motion first introduced by Mandelbrot. Hurst's rescaled range analysis indicates that the statistics of many natural phenomena are indeed best represented as fractional Brownian motion.

9.1 Brownian Motion

Robert Brown (1828) was the first to realize that the erratic motion of microscopic pollen was physical, not biological in nature as was believed before his time. Everything is subject to thermal fluctuations and molecules, macromolecules, viruses, particles and other components of the natural world are all in eternal motion with random collisions due to thermal energy. A particle at absolute temperature T has, on the average, a kinetic energy of $\frac{3}{2}kT$, where k is Boltzmann's constant. Einstein showed that this is true independent of the particle size (Einstein, 1905). Much of our understanding of thermal equilibrium and how it is reached is based on the enormous amount of research that has centered on the concept of '*Brownian motion*.' The motion of a 'Brownian particle' as seen under the microscope consists apparently of steps in a random direction and with a step-length that has some characteristic value, and therefore *random walk* is a term often used in the context of Brownian motion.

We emphasize that in Brownian motion it is not the position of the particle at one time that is independent of the position of the particle at another; it is the displacement of that particle in one time interval that is independent of the displacement of the particle during another time interval.

Increasing the resolution of the microscope and the time resolution only produces a similar random walk. As we shall see Brownian motion is self-similar. If the time axis is included as an extra dimension, the particle position as a function of time — also called the *record* of the motion — is *not* self-similar but *self-affine*. The distinction between self-similar and self-affine will be important in the following discussions.

Diffusion is best understood in terms of Brownian motion. Some of the interesting and often surprising effects of Brownian motion in biology are very nicely described in a small book by Berg (1983).

9.2   Random Walk in One Dimension

Consider a situation in which a 'particle' moves on a line — the x-axis — by jumping a step-length of either $+\xi$ or $-\xi$ every τ seconds. In modeling diffusion we consider ξ to be some microscopic length, say the particle diameter, and τ is a microscopic time — the collision time.

Rather than considering a fixed length ξ let the step-length be given by a Gaussian or normal probability distribution:

$$p(\xi, \tau) = \frac{1}{\sqrt{4\pi\mathcal{D}\tau}} \exp\left(-\frac{\xi^2}{4\mathcal{D}\tau}\right) . \tag{9.1}$$

We may envisage the random walk process on the atomistic scale as follows: At intervals τ a step-length ξ is chosen at random so that the probability of finding ξ in the range from ξ to $\xi + d\xi$ is $p(\xi, \tau)d\xi$. A sequence of such steps $\{\xi_i\}$ is a set of *independent* Gaussian random variables. The variance of the process is

$$\langle \xi^2 \rangle = \int_{-\infty}^{\infty} \xi^2 p(\xi, \tau)\, d\xi = 2\mathcal{D}\tau . \tag{9.2}$$

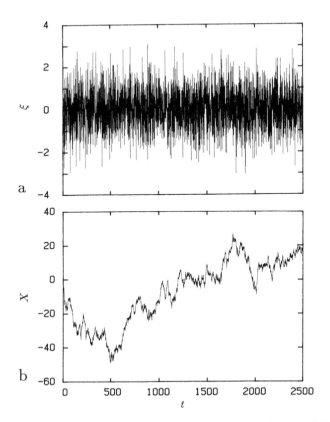

FIGURE 9.1: A sequence of independent Gaussian random variables with zero mean and unit variance. (a) The independent random steps of the 'particle.' (b) The position of the particle. The time is in units of the 'atomistic' time τ between steps.

The parameter $\mathcal{D}$ is the *diffusion coefficient*. It follows from equation (9.2) that the diffusion coefficient is given by the *Einstein relation*:

$$\mathcal{D} = \frac{1}{2\tau}\langle \xi^2 \rangle , \qquad (9.3)$$

where $\langle \xi^2 \rangle$ is the mean square jump distance. The equation (9.3) is valid under rather general conditions even for the case in which the jumps do not occur at regular intervals and when the probability distribution for the step-length ξ is discrete, is continuous or has some rather arbitrary shape.

We obtain a *normalized* Gaussian random process by the replacement $\xi \leftarrow \xi/\sqrt{2\mathcal{D}\tau}$ so that the new ξ has zero average and the variance is $\langle \xi^2 \rangle = 1$. In figure 9.1a we show a sequence of *normalized* Gaussian random

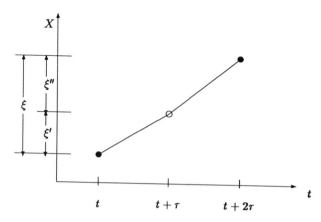

FIGURE 9.2: The increment ξ in the position of a Brownian particle at time step 2τ is the sum of two independent increments ξ' and ξ''.

variables. The sequence $\xi_1, \xi_2, \ldots$ is the sequence of steps of the random walk, whereas the *position* of the particle on the x-axis is given by

$$X(t = n\tau) = \sum_{i=1}^{n} \xi_i \, . \tag{9.4}$$

The curve in figure 9.1b shows the position of the particle as a function of time. Note, however, that the curve is really a discrete set of points — we just did not bother to lift the pen between points. In the limit of arbitrarily small time steps the random variables become a *random function* $X(t)$. The graph of the random function will look similar to figure 9.1b and is called the *record* of the random function $X(t)$. Mandelbrot calls this record a *Brown* function and denotes it by $B(t)$.

9.3   Scaling Properties of One-Dimensional Random Walks

In practice we do not observe Brownian motion with infinite resolution and we must consider the case in which we observe the particle position only at intervals $b\tau$, where b is some arbitrary number. Start by considering the particle position only every second time step, i.e., $b = 2$, as indicated in figure 9.2.   The increment ξ in the particle position is now the sum of two independent increments ξ' and ξ''. The joint probability $p(\xi'; \xi'', \tau)d\xi'd\xi''$ that the first increment ξ' is in the interval $[\xi', \xi'+d\xi')$ and that the second increment ξ'' is in the interval $[\xi'', \xi'' + d\xi''\rangle$ is given in terms of $p(\xi, \tau)$ in equation (9.1) by

$$p(\xi'; \xi'', \tau) = p(\xi', \tau)p(\xi'', \tau) .$$

The joint probability distribution is the product of the two probability densities for each of the variables because the two increments are statistically independent. The two increments must add up to the total increment ξ, and by integrating over all possible combinations of ξ' and ξ'' we find that the probability density for the increment ξ is given by

$$p(\xi, 2\tau) = \int_{-\infty}^{\infty} d\xi'\, p(\xi - \xi', \tau)p(\xi', \tau) = \frac{1}{\sqrt{4\pi \mathcal{D} 2\tau}} \exp\left(-\frac{\xi^2}{4\mathcal{D} 2\tau}\right) . \quad (9.5)$$

Thus we see that when viewed with only half the time resolution the increments of the particle position are still a Gaussian random process with $\langle \xi \rangle = 0$. However, the variance has increased: $\langle \xi^2 \rangle = 4\mathcal{D}\tau$. The argument is easily extended to time intervals $b\tau$ between observations, with the result

$$p(\xi, b\tau) = \frac{1}{\sqrt{4\pi \mathcal{D} b\tau}} \exp\left(-\frac{\xi^2}{4\mathcal{D} b\tau}\right) . \quad (9.6)$$

We therefore conclude that whatever the number b of microscopic time steps between observations, we always find that the increments in the particle position constitute an independent Gaussian random process with $\langle \xi \rangle = 0$ and a variance of

$$\langle \xi^2 \rangle = 2\mathcal{D}t \qquad \text{with} \quad t = b\tau . \quad (9.7)$$

In figure 9.3 we show the particle position as observed every fourth microscopic time step for a process of 10,000 independent increments of zero mean and unit variance, i.e., the same process shown in figure 9.1. Here each of the increments is the sum of 4 independent steps, and we see that there is little to distinguish figure 9.1a from figure 9.3a — except for the scale of the increments, which now is approximately a factor of 2 larger. Similarly, the particle position *record* in figure 9.3b is statistically the same as that in figure 9.1b — again apart from the scale used for the x-axis. However, in any given finite realization the two records will be quite different in local detail and the vertical scale will not be changed by the expected factor of $\sqrt{b}$.

The result that the Brownian record looks 'the same' under a change of resolution is called a *scale invariance* or *symmetry* of the Brownian record. This *scaling property* of Brownian motion may be expressed in an explicit way by transforming equation (9.1) by the replacements, $\hat{\xi} = b^{1/2}\xi$ and $\hat{\tau} = b\tau$, in equation (9.1), i.e., we change the time scale by a factor b, and we

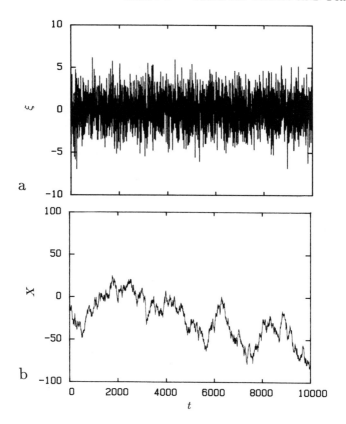

FIGURE 9.3: A sequence of independent Gaussian random variables with zero mean and unit variance — 'observed' at every fourth time step, i.e., at intervals of length 4τ. (a) The independent random steps of the 'particle.' (b) The position of the particle. The time is in units of the 'atomistic' time τ between steps.

change the length scale by a factor $b^{1/2}$. The result of this transformation in equation (9.1) is a scaling relation for the probability density:

$$p(\hat{\xi} = b^{1/2}\xi, \hat{\tau} = b\tau) = b^{-1/2}p(\xi, \tau) . \tag{9.8}$$

Here the prefactor $b^{-1/2}$ ensures that the probability density is properly normalized:

$$\int_{-\infty}^{\infty} d\hat{\xi}\, p(\hat{\xi}, \hat{\tau}) = \int_{-\infty}^{\infty} d\xi\, p(\xi, \tau) = 1 .$$

The equation (9.8) shows that the Brownian random process is *invariant in distribution* under a transformation that changes the time scale by b and the length scale by $b^{1/2}$ because it satisfies equation (9.8). As we discuss later a transformation that scales time and distance by different factors is called *affine*, and curves that reproduce themselves in some sense under an affine transformation are called *self-affine*.

The probability distribution for the particle position $X(t)$ is also found by the methods used above and we obtain the result

$$P(X(t) - X(t_0)) = \frac{1}{\sqrt{4\pi \mathcal{D}|t - t_0|}} \exp\left(-\frac{[X(t) - X(t_0)]^2}{4\mathcal{D}|t - t_0|}\right) , \qquad (9.9)$$

which satisfies the scaling relation

$$P(b^{1/2}[X(bt) - X(bt_0)]) = b^{-1/2} P(X(t) - X(t_0)) . \qquad (9.10)$$

With this probability distribution for the particle position, it follows that the average and the variance of the particle position $X(t)$ are given by

$$
\begin{aligned}
\langle X(t) - X(t_0) \rangle &= \int_{-\infty}^{\infty} \Delta X \, P(\Delta X, t - t_0) \, d\Delta X = 0 , \\
\langle [X(t) - X(t_0)]^2 \rangle &= \int_{-\infty}^{\infty} \Delta X^2 \, P(\Delta X, t - t_0) \, d\Delta X , \\
&= 2\mathcal{D}|t - t_0| ,
\end{aligned}
\qquad (9.11)
$$

where $X(t_0)$ is the particle position at some reference time t_0, and ΔX is the increment in the particle position: $\Delta X = X(t) - X(t_0)$

The position $X(t)$ of a Brownian particle is a *random function* of time t. Wiener (1923) introduced the random function for Brownian motion as follows. Consider a normalized independent Gaussian random process $\{\xi\}$. Let the increments in the position of the Brownian particle be given by

$$X(t) - X(t_0) \sim \xi |t - t_0|^H \qquad (t \geq t_0) , \qquad (9.12)$$

for *any* two times t and t_0. Here, $H = 1/2$ for ordinary Brownian motion. Equation (9.12) defines a *random function* and applies at the instant t_0 whether or not the earlier values of $X(t)$ (for $t \leq t_0$) are known. Often equation (9.12) is supplemented with the extra condition $X(0) = 0$, but this is only a matter of convenience. With the definition in equation (9.12) one finds the position $X(t)$ given the position $X(t_0)$ by choosing a random number ξ from a Gaussian distribution, multiplying it by the time increment $|t - t_0|$ and adding the result to the given position $X(t_0)$. This procedure is valid also for $t < t_0$. The function defined by equation (9.12) is continuous but it has no derivative. It follows from the definition (9.12) that the random function $X(t)$ has the distribution given in equation (9.9).

It follows from equation (9.9) that the reduced variable x defined by

$$x = \frac{X(t) - X(t_0)}{\sqrt{2\mathcal{D}\tau}(|t - t_0|/\tau)^H} \qquad (9.13)$$

has for all t and t_0 a Gaussian probability distribution:

$$p(x) = \frac{1}{\sqrt{2\pi}} \exp\left(-\frac{1}{2}x^2\right) \tag{9.14}$$

with zero mean and unit variance.

9.4 Fractional Brownian Motion

Mandelbrot has introduced the concept of *fractional Brownian motion* as a generalization of the random function $X(t)$ by changing the exponent from $H = 1/2$ to any real number in the range $0 < H < 1$ in equations (9.12) and (9.13) (Mandelbrot and Van Ness, 1968; Mandelbrot, 1982), and he denotes such functions $B_H(t)$. Cases where $H \neq 1/2$ are properly fractional, whereas the case $H = 1/2$ is the special case of *independent* increments valid for Brownian motion, and for this case we write $B(t) = B_{1/2}(t)$.

Using $B_H(t)$ instead of $X(t)$ for the particle position it follows from equations (9.13) and (9.14) that a fractional Brownian process has zero average increments:

$$\langle B_H(t) - B_H(t_0)\rangle = 0 \,,$$

and a *variance of increments* $V(t - t_0)$, given by

$$\begin{aligned} V(t - t_0) &= \langle [B_H(t) - B_H(t_0)]^2 \rangle \\ &= 2\mathcal{D}\tau(|(t - t_0)/\tau|)^{2H} \sim |t - t_0|^{2H} \,. \end{aligned} \tag{9.15}$$

It is apparent that both ordinary and fractal Brownian motion have variances that diverge with time.

It is important to realize that fractional Brownian motion has infinitely long-run correlations. In particular past increments are correlated with future increments: Given the increment $B_H(0) - B_H(-t)$ from time $-t$ to 0 the probability of having an increment $B_H(t) - B_H(0)$ averaged over the distribution of the past increments is

$$\langle [B_H(0) - B_H(-t)][B_H(t) - B_H(0)]\rangle$$

For convenience set $B_H(0) = 0$ and use units so that $\tau = 1$ and $2\mathcal{D}\tau = 1$. The correlation function of future increments $B_H(t)$ with past increments $-B_H(-t)$ may be written

$$C(t) = \frac{\langle -B_H(-t)B_H(t)\rangle}{\langle B_H(t)^2\rangle} = 2^{2H-1} - 1 \,, \tag{9.16}$$

where we have normalized with the variance of B_H. The last equation follows directly from equation (9.15) .

We first note that for $H = 1/2$ we find that the correlation of *past* and *future* increments $C(t)$ vanishes for all t — as is required for an *independent* random process. However, for $H \neq 1/2$ we have $C(t) \neq 0$, *independent* of t! This is a remarkable feature of fractional Brownian motion which leads to *persistence* or *antipersistence*. For $H > 1/2$, we have *persistence*. In this case, if we for some time in the past have a positive increment — i.e., an increase — then we also have on the average an increase in the future. Therefore an increasing trend in the past implies an increasing trend in the future for processes with $H > 1/2$ — and furthermore this applies for arbitrarily large t! Conversely a decreasing trend in the past implies on the average a continued decrease in the future.

Antipersistence is the term for the case $H < 1/2$. In this case an increasing trend in the past implies a decreasing trend in the future, and a decreasing trend in the past makes an increasing trend in the future probable.

It should be noted that the behavior of a statistical record given by equation (9.16) is in conflict with what is normally either assumed or proven for statistical records and for physical systems. In fact for statistical physics the underlying assumption used to be that events may be correlated when separated in time Δt but they will definitely become uncorrelated in the limit $\Delta t \rightarrow \infty$. This statistical independence at large time and/or space separations is an essential ingredient of the concept of thermal equilibrium. There are exceptions: As a second-order phase transition point, for example the critical point of a fluid, is approached, the correlation functions of the density develop a component which has no intrinsic length scale or time scale. As a consequence the free energy of the system has a critical part that has the *scaling* form of equation (2.13), and power-law behavior of the correlation functions becomes the rule rather than the exceptional case.

For time series of observations we have discussed a method, developed by Hurst, which when applied to many natural phenomena leads to the conclusion that they appear to exhibit persistence over a wide range of time scales. Fractional Brownian motion is useful in modeling the time series of these phenomena.

9.5   Definition of Fractional Brownian Motion

Insight into the nature of fractional Brownian motion may be obtained by implementing such a process by computer simulation to generate results similar to those presented for ordinary Brownian motion in figure 9.1. Mandelbrot and Van Ness (1968) defined the random function $B_H(t)$ with zero mean roughly as follows:

$$B_H(t) = \frac{1}{\Gamma(H + \frac{1}{2})} \int_{-\infty}^{t} (t - t')^{H-1/2} \, dB(t') \, . \tag{9.17}$$

Here $\Gamma(x)$ is the gamma function. This definition states that the value of the random function at time t depends on all previous increments $dB(t')$ at time $t' < t$ of an ordinary Gaussian random process $B(t)$ with average zero and unit variance.

The notation $dB(t)$ for a random variable becomes transparent when one tries to evaluate the integral by replacing it by a summation. Choose time units so that t is an integer time variable, and divide each time unit into n small time steps for the purpose of approximating the integral by a sum. Then we may write the time of integration $t' = i/n$ with $i = -\infty, \ldots, -2/n, -1/n, 0, 1/n, \ldots, t/n$. The increment $dB(t')$ of the underlying independent Gaussian process may then be written $n^{-1/2}\xi_i$, where ξ_i now is a discrete Gaussian random variable of average zero and unit variance. The factor $n^{-1/2}$ in front of ξ takes care of the rescaling of Brownian increments with decreasing time steps [see equation (9.8)]. We therefore obtain the approximate expression

$$B_H(t) \simeq \frac{1}{\Gamma(H + \frac{1}{2})} \sum_{i=-\infty}^{nt} \left(t - \frac{i}{n} \right)^{H-1/2} n^{-1/2} \xi_i \, . \tag{9.18}$$

It is clear that this sum does not exist and that the integral in equation (9.17) is divergent as $t' \to -\infty$. We have to replace the rough definition by a more precise definition used by Mandelbrot and Van Ness (1968). Given the value $B_H(t = 0)$ we have

$$B_H(t) - B_H(0) = \frac{1}{\Gamma(H + \frac{1}{2})} \int_{-\infty}^{t} K(t - t') \, dB(t') \, . \tag{9.19}$$

Here the simple power-law kernel in equation (9.17) is replaced by the modified kernel

$$K(t - t') = \begin{cases} (t - t')^{H-1/2} \, , & 0 \le t' \le t \\ \{(t - t')^{H-1/2} - (-t')^{H-1/2}\} \, , & t' < 0 \end{cases} \, . \tag{9.20}$$

This kernel vanishes quickly enough as $t' \to -\infty$ to make the expression properly define the random function $B_H(t)$.

The equation (9.19) has the form of a general linear response expression. Here an independent Gaussian increment $dB(t')$ of magnitude unity at time t' gives a contribution to the fractal Brownian 'particle' position $B_H(t)$ at a later time t as determined by the linear response function $K(t - t')$.

The unusual feature of $K(t) \sim t^{H-1/2}$, is that the power-law form has no intrinsic time scale, or unit of time, and we find the *scaling form* of equation (9.19) by changing the time scale by a factor b to obtain

$$B_H(bt) - B_H(0) = \frac{1}{\Gamma(H + \frac{1}{2})} \int_{-\infty}^{bt} K(bt - t') \, dB(t') . \qquad (9.21)$$

Here we introduce a new integration variable $t' = b\hat{t}$ and use the result that for an independent Gaussian process we have *in distribution* that $dB(t' = b\hat{t}) = b^{1/2} dB(\hat{t})$. Using the relation $K(bt - b\hat{t}) = b^{H-1/2} K(t - \hat{t})$ we then find that

$$B_H(bt) - B_H(0) = b^H \{B_H(t) - B_H(0)\} \qquad (9.22)$$

is valid in distribution for *any* value of b. In particular, we may choose $t = 1$ and $\Delta t = bt$ and conclude that the increment of the fractal Brownian 'particle' position,

$$B_H(\Delta t) - B_H(0) = |\Delta t|^H \{B_H(1) - B_H(0)\} \sim |\Delta t|^H , \qquad (9.23)$$

is proportional to $|\Delta t|^H$ *in distribution*. It follows therefore that the variance of increments is given by equation (9.15) with $t_0 = 0$ and $\Delta t = t - t_0$. This result is of course the basis for choosing the definition (9.19) for B_H in the first place.

9.6   Simulation of Fractional Brownian Motion

The discrete version of $B_H(t)$ given in equation (9.18) has to be modified with the proper kernel in order to make the sum convergent. However, any calculation of B_H must use a finite number of terms and the sums can only cover a range M in the integer time t. Dividing each integer time step into n intervals for the purpose of approximating the integral we have the following approximation (Mandelbrot and Wallis, 1969b–d):

$$B_H(t) - B_H(t - 1) = \frac{1}{\Gamma(H + \frac{1}{2})} \sum_{i=n(t-M)}^{nt} K(t - \tfrac{i}{n}) n^{-1/2} \xi_i . \qquad (9.24)$$

Here $\{\xi_i\}$, with $i = 1, 2, \ldots, M, \ldots$, is a set of Gaussian random variables with unit variance and zero mean. The kernel K is given by equation (9.20). With a change of the summation variable and a rearrangement of terms we find that the discrete fractional Brownian increments are given by

$$
B_H(t) - B_H(t-1) = \frac{n^{-H}}{\Gamma(H + \frac{1}{2})} \left\{ \sum_{i=1}^{nt} (i)^{H-1/2} \xi_{(1+n(M+t)-i)} \right.
$$
$$
\left. + \sum_{i=1}^{n(M-1)} \left((n+i)^{H-1/2} - (i)^{H-1/2} \right) \xi_{(1+n(M-1+t)-i)} \right\} \tag{9.25}
$$

With the procedure given in equation (9.25) a sequence of *increments* in B_H may be generated from a sequence of Gaussian random variables. Note that this approximation is a sliding average over the Gaussian process with a power-law weight function. Since only M integer time steps are included in the summation it follows that for integer times $t \gg M$, the increments will become independent and this approximation to B_H becomes an independent Gaussian process. It is clear that the algorithm given in equation (9.25) is inefficient since the the sums of nM terms have to be evaluated for each increment in B_H. Mandelbrot (1971) has proposed a fast algorithm for fractional Gaussian noise based on a weighted sum of a series of Markov-Gauss random variables with increasing correlation times in addition to a high-frequency Markov-Gauss term. However, the present algorithm suits the present exposition better and we will content ourselves with simulations having moderately large M.

The effect of increasing n is to give a more precise approximation to the short-term behavior of $B_H(t)$ (which is not very important here) and we have chosen $n = 8$ in the illustrations.

We have evaluated B_H from the Gaussian process of 27,500 independent steps for which the first 2500 steps are shown in figure 9.1. In figure 9.4 we show the fractal noise, i.e., the increments of B_H given by $\Delta B_H(t) = B_H(t) - B_H(t-1)$. For ordinary Brownian increments $H = 1/2$, and the noise is an independent Gaussian process which is what is normally called *white noise*. In figure 9.4 the fractal noises shown have $H = 0.7$ and $H = 0.9$. There is no dramatic change in the noise as H is increased. A closer inspection reveals, however, that as H increases the low-frequency noise increases and generates large amplitude excursions compared with the high-frequency components.

In figure 9.5 we show the fractal Brownian function $B_H(t)$ as a function of time, using $B_H(0) = 0$. This function corresponds to the position of a particle that starts at the origin and uses the increments in figure 9.4 as steps along the x-axis. As the exponent H is increased the record of the

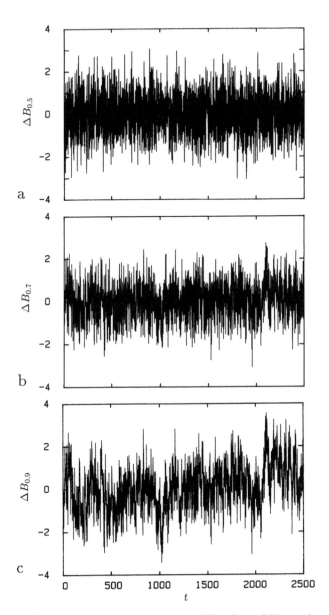

FIGURE 9.4: Fractal noise or increments of the fractal Brownian function B_H evaluated with $M = 700$, $n = 8$. (a) Ordinary Brownian increments for $H = 1/2$. (b) Fractional increments for $H = 0.7$. (c) Fractional increments for $H = 0.9$.

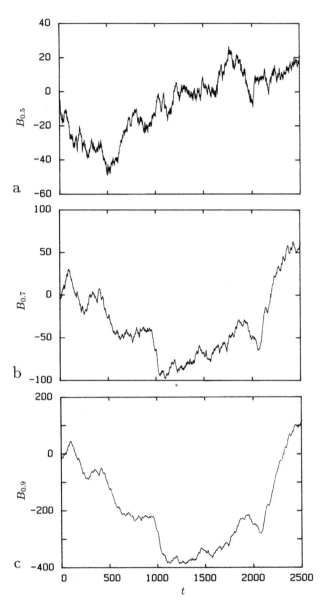

FIGURE 9.5:   The fractal Brownian function B_H evaluated with $M = 700$, $n = 8$ and with $B_H(0) = 0$. (a) The ordinary Brownian function for $H = 1/2$. (b) Fractional Brownian function for $H = 0.7$. (c) Fractional Brownian function for $H = 0.9$.

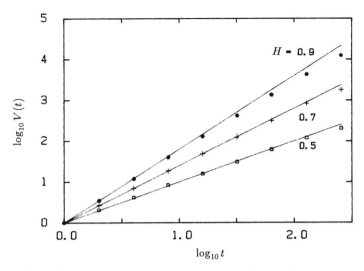

FIGURE 9.6: The correlation of increments for fractal Brownian function $V(t)$ evaluated with $M = 700$, $n = 8$ and with $B_H(0) = 0$. The lines correspond to $V(t) = |t|^{2H}$, with $H = 0.5$, 0.7, 0.9.

particle position increases in amplitude, and the noise is proportionally reduced.

As compared with Brownian motion, fractal motion with $H > 1/2$ moves anomalously large distances from the origin. In fact fractal Brownian motion has a variance in position given by equation (9.15), so if we use the Einstein relation in the form of equation (9.11) we can define an anomalous diffusivity for fractal diffusion:

$$\mathcal{D}_H = \frac{1}{2}\frac{d}{dt}\langle X(t)^2 \rangle = \mathcal{D}|t|^{2H-1} . \qquad (9.26)$$

The anomalous diffusivity is an important concept in the discussion of fractal transport phenomena. It arises in many contexts, for example in the discussion of the electrical conductivity of random systems. It should be noted that the anomalous character of the diffusivity in equation (9.26) is due to the fractal nature of a random walk in Euclidean space. If the walk is restricted to a fractal set embedded in Euclidean space one still finds an anomalous diffusivity but with changed exponent in the power-law dependence on time (see for instance Gefen et al., 1983; Aharony, 1985; Stanley, 1985).

The normalized variance of increments as a function of the lag-time t, given by equation (9.15), may be written

$$V(t) = \langle [B_H(t) - B_H(0)]^2 \rangle / \langle B_H(t)^2 \rangle = |t|^{2H} . \qquad (9.27)$$

We estimate $V(t)$ from the records shown in figure 9.5, and find that the resulting $V(t)$ quite nicely approximates the expected behavior given in equation (9.27) for various values of H. However, the effect of using a finite value of M, and a finite length of the underlying independent Brownian motion, reflects itself by the fact that the points estimated from the simulation fall below the theoretical expectation for lags of the order $t \geq M$ (see figure 9.6). Therefore at $t \sim M$ the approximate discrete fractal noises begin to cross over to independent white noise. We may extend arbitrarily the domain over which the simulated noise is fractional by increasing M. However, more efficient algorithms must be used if one needs fractional noise over a very large range of time-lags.

9.7 R/S Analysis of Fractional Brownian Motion

The scaling form of equation (9.22) gave the result (9.23) that the random function $B_H(\Delta t)$ is proportional to $|\Delta t|^H$. This has the consequence (Mandelbrot and Wallis, 1969d; Mandelbrot, 1982) that the range $R(\tau)$ with lag τ is also a random function with the scaling property

$$R(\tau) \sim \tau^H .$$

Since the true variance $S = 1$ and the sample variance is $\simeq 1$ for the normalized fractal Brownian function, it follows that the rescaled range R/S is given by

$$R(\tau)/S \sim \tau^H , \tag{9.28}$$

in distribution. We therefore find that the Hurst exponent H can be *estimated* by a fit of equation (9.28) to experimental or simulated results.

We have performed an R/S analysis of our simulations B_H, as a test of this relation. In figure 9.7 we show the R/S analysis of the independent Gaussian process shown in figure 9.1. The Hurst exponent estimated from our simulations of this process is $H = 0.510 \pm 0.008$, which is in good agreement with the theoretical result $H = \frac{1}{2}$.

For the fractal Brownian function $B_{0.9}(t)$, shown in figure 9.4c, the R/S analysis is shown in figure 9.8. We see that these results are quite different from the ordinary Brownian process. However, we find that the Hurst exponent $H = 0.81 \pm 0.02$ estimated from the data is a little below the $H = 0.9$ used to generate the data. It should be remembered, however, that the simulations generate only *approximate* fractal Brownian functions since we use a finite memory term $M = 700$, and a finite resolution $n = 8$. It is therefore reasonable that the Hurst exponent estimated from the

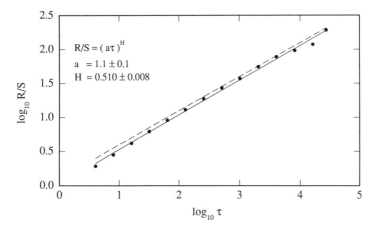

FIGURE 9.7: R/S as a function of the lag τ for a normalized independent Gaussian variable ξ. The dotted line is the asymptotic theoretical expectation $R/S = \sqrt{\pi\tau/2}$. The line is the fitted curve $R/S = (a\tau)^H$ with $H = 0.510 \pm 0.008$ and $a = 1.1 \pm 0.1$.

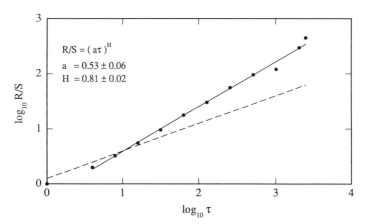

FIGURE 9.8: R/S as a function of the lag τ for the fractal Brownian function $B_H(t)$ with $H = 0.9$. The dotted line is the asymptotic result for an independent Gaussian process $R/S = \sqrt{\pi\tau/2}$. The line is the fitted curve $R/S = (a\tau)^H$ with $H = 0.81 \pm 0.02$ and $a = 0.53 \pm 0.06$.

simulations is a little low, since at lags $\tau > 700$ we start to cross over to an independent Gaussian process with $H = 1/2$. We conclude that the Hurst exponent may be accurately obtained from an analysis of well-defined data sets with about 2500 observations.

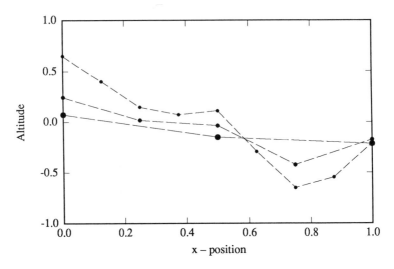

FIGURE 9.9: The process of successive random additions. The lines connect points of the same generation. The three largest circles represent the first random additions to three altitudes that initially were zero. The five next largest circles are obtained by interpolation of the first three values at the midpoints, and then adding smaller random increments to all five points. The next generation produces the values at nine points, marked with the smallest dots.

9.8   Successive Random Addition

R. F. Voss has introduced methods for generating not only ordinary fractional Brownian motion but also fractionally Brownian surfaces and clouds. Voss (1985b) calls his algorithm *successive random addition*. We may describe his algorithm as follows, referring to figure 9.9. In order to construct a fractional Brownian curve where altitude or vertical position $X(t)$ is a fractional Brownian motion we must require that the variance of increments for the position be given by

$$V(t) = \langle [X(t) - X(0)]^2 \rangle = |t|^{2H} \sigma_0^2 . \tag{9.29}$$

This is just a rewritten form of equation (9.27), where σ_0^2 is the (initial) variance of the random additions to be discussed.

   The following process, introduced by Voss, generates a fractional Brownian motion to an arbitrary resolution. The starting point is a sequence of positions $X(t_1), X(t_2), \ldots, X(t_N)$ at the times $t_1, \ldots, t_N$. We choose $N = 3$ at $t_i = 0, 1/2, 1$ and set the positions equal to zero. Next, the positions $X(t_1), X(t_2), X(t_3)$ are given random additions chosen from a normal distribution with zero mean and unit variance: $\sigma^2 = \sigma_1^2 = 1$. The midpoints of the time intervals become additional times at which the positions are estimated by interpolation. The times are now $t_1, \ldots, t_5 = 0, 1/4, 1/2, 3/4, 1$. All positions are now given a random addition with zero mean and a *reduced* variance

$$\sigma_2^2 = (1/2)^{2H} \sigma_1^2 \; .$$

These five new positions are again interpolated to the midpoints of the time intervals to give nine positions at nine times. After n applications of this algorithm we have defined the position of the fractional Brownian particle at $(1 + 2^n)$ times. The positions are obtained by the interpolation and random addition process. The variance of the addition in the n-th generation of this process is

$$\sigma_n^2 = (1/2)^{2H} \sigma_{n-1}^2 = (1/2)^{2H\,n} \sigma_0^2 \; .$$

Voss shows that this process leads to self-affine curves, which have a fractal dimension $D = 2 - H$.

We have generated such curves for various values for the Hurst exponent and the results are shown in figure 9.10. The curves have been rescaled so that they all have zero mean and unit sample variance. The case $H = 0.5$ is that of ordinary Brownian motion, found in many applications. The noise from an amplifier, for example, is often assumed to have the form given by the $H = 0.5$ curve.

PERSISTENT behavior is obtained for $0.5 < H < 1$. An example of a persistent process with $H = 0.92$ is the statistics of ocean waves discussed in chapter 11. Persistence means here that if the wave-height has been increasing for a period t, then it is expected to continue to increase for a similar period. Conversely, if the wave-height is observed to decrease for a period t, then it is expected to continue to decrease for a similar period. In other words, persistent stochastic processes exhibit rather clear trends with relatively little noise. In fact one tends to look for periodicities in the records of persistent stochastic processes (see figure 9.10).

ANTI-PERSISTENT stochastic processes, on the other hand, tend to show a decrease in values following previous increases, and show increases following previous decreases. Fractal Brownian processes exhibiting such behavior have H in the range 0 to $1/2$. The record of an anti-persistent process, such as the $H = 0.1$ curve shown in figure 9.10, appears very 'noisy.' They have local noise of the same order of magnitude as the total excursions of the record.

Voss has generalized the successive random addition algorithm to two and higher dimensions. His fractal landscapes are generated by interpolation on a square grid and with the same random addition with decreasing

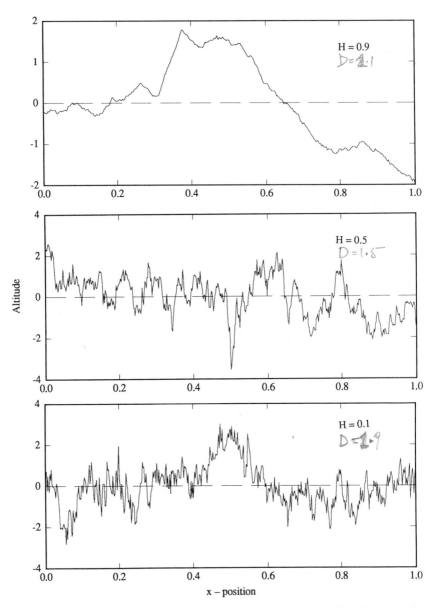

FIGURE 9.10: Fractional Brownian curves generated using Voss's successive random addition algorithm with different values for the Hurst exponent H. The fractal dimension of the curves is $D = 2 - H$. The curves have been calculated with a resolution of 1/2048.

variance as discussed above. We show examples of landscapes generated using Voss's algorithm in chapter 13.

Voss (1985b) also shows that the *lacunarity* of the resulting fractal surfaces can be controlled by choosing the reduction ratio to be r instead of $1/2$, so that in the n-th generation one gives random additions with variance $\sigma_n^2 = r^{2H}\sigma_0^2$. Voss has generated pictures of clouds by assigning water concentrations $c(\mathbf{x})$ in three-dimensional space and painting regions that have c above some fixed level white. His cloud pictures have a quality comparable to that obtained by the best painters. In this view clouds are fractal volumes in a four-dimensional self-affine space consisting of the usual three spatial coordinates plus the dimension of water concentration.

Chapter 10

Self-Similarity and Self-Affinity

The probability distribution for Brownian motion satisfies the *scaling relations* given by equations (9.8) and (9.10), analogous to the scaling relations (2.12), (2.13) and (2.16) discussed previously. However, there is a very important extension here. The first step is that we now have a function that is scaling in two variables ξ and t. This is nothing new since the von Koch curve in figure 2.8 already depends on two variables x and y, and we have already shown that the curve is self-similar with a scaling factor r that is directly related to the fractal dimension D of the curve — see equation (2.10). The second and important step is that time and position are scaled with *different* ratios — when we scale time by b to bt, we scale position by b^H.

The similarity transformation discussed (see chapter 2) transforms points $\mathbf{x} = (x_1, \ldots, x_E)$ in E-dimensional space into new points $\mathbf{x}' = (rx_1, \ldots, rx_E)$ with the *same* value of the scaling ratio r. A bounded *self-similar* fractal set of points $\mathcal{S}$ is *self-similar* with respect to a *scaling ratio* r if $\mathcal{S}$ is the *union* of N non-overlapping subsets $\mathcal{S}_1, \ldots, \mathcal{S}_N$, *each* of which is congruent to the set $r(\mathcal{S})$ obtained from $\mathcal{S}$ by the similarity transform defined by $0 < r < 1$. Here *congruent* means that the set of points $\mathcal{S}_i$ is identical to the set of points $r(\mathcal{S})$ after possible translations and/or rotations of the set. The similarity dimension is then given by

$$D_S = \frac{\ln N}{\ln 1/r} \, . \tag{10.1}$$

The set $\mathcal{S}$ is *statistically self-similar* when $\mathcal{S}$ is the union of N distinct subsets each of which is scaled down by r from the original and is identical in all statistical respects to $r(\mathcal{S})$. Often, random sets — such as a coastline — are statistically self-similar not only for a given value of the scaling ratio r, but for all scaling ratios above some lower cutoff (the micro-scale) and

184

some upper cutoff (the macro-scale). For such sets the *box-counting* method is useful in estimating the fractal dimension of the set and it coincides with D_S.

In many cases of interest we study sets that are not self-similar. For instance, when we study the motion of a Brownian particle the position of the particle and the time are different physical quantities and we cannot expect X and t to scale with the same ratio. We therefore need to discuss the concepts related to *self-affinity*.

An *affine transformation* transforms a point $\mathbf{x} = (x_1, \ldots, x_E)$ into new points $\mathbf{x}' = (r_1 x_1, \ldots, r_E x_E)$, where the scaling ratios $r_1, \ldots, r_E$ are *not* all equal. For an example see the following section.

A bounded set S is *self-affine* with respect to a *ratio vector* $\mathbf{r} = (r_1, \ldots, r_E)$ if S is the *union* of N nonoverlapping subsets $S_1, \ldots, S_N$, *each* of which is congruent to the set $\mathbf{r}(S)$ obtained from S by the affine transform defined by $\mathbf{r}$. Here *congruent* means that the set of points S_i is identical to the set of points $\mathbf{r}(S)$ after possible translations and/or rotations of the set.

The set S is *statistically self-affine* when S is the union of N nonoverlapping subsets each of which is scaled down by $\mathbf{r}$ from the original and is identical in all statistical respects to $\mathbf{r}(S)$

More details on self-affine and self-similar sets are found in Mandelbrot's book (1982), and in papers by Voss (1985a,b). A paper by Barnsley and Sloan (1988) discusses interesting applications of the iteration of self-affine transformations.

The fractal dimension of even the simplest self-affine fractals is not uniquely defined. For a recent discussion see Mandelbrot (1985, 1986). First, the similarity dimension is simply not defined; it exists only for self-similar fractals. What about the *box dimension* D_B? This dimension can at least be evaluated 'mechanically' for a set of points such as the *record* of a fractal Brownian function $B_H(t)$. For the independent Gaussian process we find a record of position $X(t)$ as a function of time as shown in figure 9.1. Let us cover that record with 'boxes' of width $b\tau$ in time, and of length ba in position, so that the smallest box is $\tau \times a$. The box dimension is then defined as in equation (2.4) by

$$N(b; a, \tau) \sim b^{-D_B} , \qquad (10.2)$$

where $N(b; a, \tau)$ is the number of boxes needed to cover the record. Since we have used integer time in our simulations we take the minimum box width to be $\tau = 1$, and use a small height $a = 0.0001$. We find that N indeed follows equation (10.2) as shown in figure 10.1a. From a fit of equation (10.2) to these results we find $D_B = 1.51 \pm 0.02$. If we use $a = 1$,

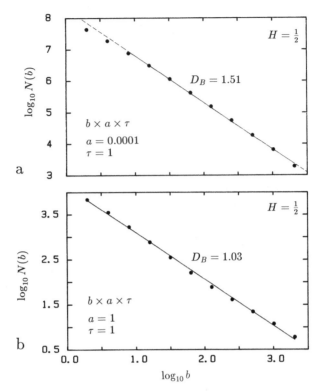

FIGURE 10.1: The number of boxes $N(b; a, \tau)$ of size $b \times (a, \tau)$ as a function of the box scale factor b for the record $B(t)$ of an independent Gaussian process consisting of 7500 steps. (a) Using $a = 0.0001$ and $\tau = 1$ we find a fit with $D_B = 1.51 \pm 0.02$. (b) Using $a = 1$ and $\tau = 1$ we find $D_B = 1.03 \pm 0.02$.

i.e., we make the minimum box height equal to the typical step length, then we find $D_B = 1.03 \pm 0.02$ (figure 10.1b).

Why do we get this difference? Let the time-span of the record be T. Then we need $T/b\tau$ segments of length $b\tau$ to cover the time axis. In each segment the range of the record is of the order $\Delta B_H(b\tau) = b^H \Delta B_H(\tau)$, and we need a stack of $b^H \Delta B_H(\tau)/ba$ boxes of height ba to cover that range. Therefore we need on the order of

$$N(b; a, \tau) = \frac{b^H \Delta B_H(\tau)}{ba} \times \frac{T}{b\tau} \sim b^{H-2} \sim b^{-D_B} \tag{10.3}$$

boxes to cover the set. We therefore find the relation

$$D = 2 - H , \qquad \textit{for self-affine records.} \tag{10.4}$$

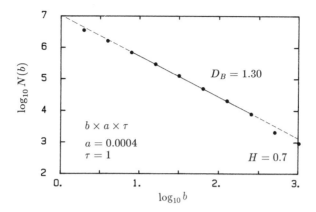

FIGURE 10.2: The number of boxes $N(b; a, \tau)$ of size $b \times (4 \cdot 10^{-4}, 1)$ as a function of the box scale factor b for the record $B_H(t)$ of an $H = 0.7$ fractal Brownian function of 2500 steps. The curve is a fit with $D_B = 1.30 \pm 0.02$.

In this argument we used boxes that were small with respect to both the length of the record T and the range of the record; therefore the relation $D = 2 - H$ holds in the high-resolution or *local* limit in the analysis of the structure of the record of a fractal function. Since we have $H = 1/2$ for independent Gaussian Brownian motion, we expect $D = 1.5$, which is consistent with the results shown in figure 10.1a. Also for a fractional Brownian function with $H = 0.7$, we find that equation (10.4) is satisfied, as shown in figure 10.2.

The argument just given breaks down if in the box-counting method we use boxes that are not small with respect to the range of the record. In particular, if we choose a to be of the order of the typical step length $a = \sqrt{\langle \xi^2 \rangle} = 1$, then in each time segment of order $b\tau$ only of the order of 1 box in the stack is needed to cover the span $\Delta B_H(b\tau)$, and therefore we find

$$N(b; a, \tau) \sim 1 \times \frac{T}{b\tau} \sim b^{-1} . \qquad (10.5)$$

The box dimension then is $D_B = 1$. This is precisely the result we found in figure 10.1b using $a = 1$. Note that we will always reach this limit if we keep increasing our box size b. By increasing b sufficiently we sooner or later reach a box size that covers the entire range of B_H, with a finite time $b\tau$. Now extend the time span T of the record to nT. The number of boxes needed to cover the full record will increase as n, which implies $D = 1$. The self-affine record has a *global* value of the fractal dimension $D = 1$, i.e., globally a self-affine record is not fractal.

We can conclude that for self-affine fractal records we must distinguish between the *local* fractal dimension $D = 2 - H$ and the *global* fractal dimension $D = 1$ (Mandelbrot, 1985, 1986; Voss, 1985a).

Another dimension that may be evaluated mechanically is the *divider dimension* obtained by walking a yardstick or divider with an opening δ along a curve to measure its length. For self-similar fractal curves, such as coastlines, we expect

$$L \sim \delta^{1-D} \; . \tag{10.6}$$

However, a yardstick that has dimension *time* when used along the t-axis and *length* when used along the x-axis, as required for the evaluation of the length of a self-affine record, does not make much sense. We can make a trace of the self-affine function on 'paper' and measure its length there. The measured length with a ruler of length δ, placed such that it covers a time step $b\tau$, will give a contribution to the length of the order of

$$\delta = (b^2\tau^2 + b^{2H}[\Delta B_H(\tau)/a]^2)^{1/2} \; . \tag{10.7}$$

By choosing a sufficiently large magnification along the x-axis, or using a small a, the last term in the square root above dominates and we have $\delta \sim b^H$. The number of such segments along the time axis is $T/b\tau \sim b^{-1} \sim \delta^{-1/H}$, and we find the length

$$L \sim \delta^{1-1/H} \sim \delta^{1-D_D} \; . \tag{10.8}$$

Therefore we find that the *divider dimension* is $D_D = 1/H$, in the local limit where the x-scale is magnified. If we instead magnify the time scale, so that the fluctuations in $X(t)$ are barely visible on the 'graph paper,' then the first term in the square root dominates and $\delta \sim b$. The length will be $L \sim \delta \times T/b \sim \delta^0$ and we find $D = 1$. Again the *global* dimension is $D = 1$, and the *local* dimension is *fractal* and $D = 1/H$. This dimension is sometimes called the *latent* fractal dimension (Mandelbrot, 1983; Voss, 1985a), and it is related to the fractal dimension of the *trail* of the Brownian particle.

We summarize the various fractal dimensions discussed here in a table.

Dimensions for the fractal Brownian function B_H			
Dimension		Trail self-similar	Record self-affine	
			Local	Global
Hausdorff	D	$1/H$	$2 - H$	–
Box	D_B	$1/H$	$2 - H$	1
Divider	D_D	—	$1/H$	1
Random walk	D_W	$1/H$	—	—
Similarity	D_S	$\frac{\ln N}{\ln 1/r} = 1/H$	—	—

10.1 The Strategy of Bold Play

An interesting example of a singular self-affine function provided by gambling theory has been discussed in a popular exposition by Billingsley (1983) of ideas introduced by Dubins and Savage (1960). He begins with the following problem: A gambler enters the casino with capital of $900 and the intent (or hope) of increasing it to $1000. He stakes $1 on each turn of the wheel, and on each turn the probability of winning $1 is p and the probability of losing $1 is $q = 1 - p$. His strategy is to play until his fortune has either increased to $1000 or dwindled to nothing. We shall see that this strategy, which I call *timid play*, is not a good strategy, and that the strategy of *bold play* defined later is much better.

The probability $M_T(x)$ of success under timid play, that is, the probability of reaching the goal $G = \$1000$, starting with initial capital $C = \$900 = xG$, is (e.g., Feller, 1968)

$$M_T(x) = \begin{cases} x \, , & \text{if } p = q = \frac{1}{2} \, , \\ \dfrac{(q/p)^{xG} - 1}{(q/p)^G - 1} \simeq \left(\dfrac{p}{q}\right)^{(1-x)G} \, , & \text{if } p \le \frac{1}{2} \le q \, . \end{cases} \tag{10.9}$$

Here the approximation is valid because $q/p > 1$ and C and G are large numbers. The small stake, compared to the distance to the goal, makes it a good approximation to consider the gambler's capital to increase and decrease by a random walk process. For fair odds the process is simply an unbiased random walk and the gambler has a passable chance of success: $M_T(x = 0.9, p = 0.5) = 0.9$. But no real casino gives fair odds. Of the 38 spaces on an ordinary roulette wheel 18 are red, 18 are black, and 2 are green and therefore $p = \frac{18}{38}$. A gambler betting on red has a probability of successfully increasing $900 to $1000 by wagering $1 at each turn of the wheel that is only about 0.00003.

Suppose the gambler is a true optimist and hopes to convert initial capital of \$100 into a final goal of \$20,000 before running out of money. His chances of success are 0.005 if $p = 0.5$ but only about 3×10^{-911} for $p = \frac{18}{38}$. The gambler has to be desperate to hope for success under these conditions. The chance of success is completely negligible. The gambler is not, however, forced to set his wager at \$1. Suppose he wagers \$10 instead. Then his initial capital consists of ten \$10 bills, and he aims to increase it to 2000 \$10 bills. This corresponds to setting $G = 2000$ and $C = 10$ in equation (10.9). The probability of success increases from 3×10^{-911} to about $(p/q)^{2000-10} \simeq 10^{-91}$. This strategy of using larger bets increases the gambler's chance for success by an enormous factor — his probability of success remains, however, completely negligible.

Large bets increase the gambler's chance of success. This leads to the strategy of *bold play* (e.g., Billingsley, 1983): *On each turn of the roulette the gambler stakes his entire current fortune if it does not exceed half the goal, and otherwise he bets the difference between the goal and the current fortune.* Bold play *maximizes* the chance of success in reaching the goal exactly when $p \leq \frac{1}{2}$. The probability of reaching the goal of $G = \$20,000$, starting with \$100, is 0.003 when $p = \frac{18}{38}$. We see that this is a very much better strategy than the strategy of timid play just discussed.

In analyzing bold play it is convenient to go to a new scale in which the gambler's fortune x lies between 0 and 1 and his goal is 1. We write $M(x)$ for the probability of success (under bold play) starting from an initial capital of x. The strategy of bold play may then be formulated as follows. Suppose first that the gambler's fortune is in the range $0 \leq x \leq \frac{1}{2}$, so that he stakes the amount x. If his fortune is to reach 1, he must win in the first turn (probability p) and then, from his new fortune $x + x = 2x$, he must go on to eventual success (probability $M(2x)$). The product $pM(2x)$ is the probability of winning on the first turn and then going on to ultimate success. Therefore the probability of success starting with $x \leq \frac{1}{2}$ is given by the first of the following two equations:

$$M(x) = p\,M(2x)\,, \qquad \text{for } 0 \leq x \leq \tfrac{1}{2}\,,$$
$$M(x) = p + (1 - p)\,M(2x - 1)\,, \quad \text{for } \tfrac{1}{2} \leq x \leq 1\,. \tag{10.10}$$

Suppose next that the gambler starts with an initial fortune in the range $\frac{1}{2} \leq x \leq 1$, so that he stakes the amount $1 - x$. He can win on the first turn of the wheel (probability p). He can also lose on the first turn (probability $q = 1 - p$), and then go on to win starting with the new capital $x - (1 - x) = 2x - 1$ (probability $M(2x - 1)$). The probability of success starting with capital $\frac{1}{2} \leq x \leq 1$ is therefore given by the second of the pair of equations (10.10).

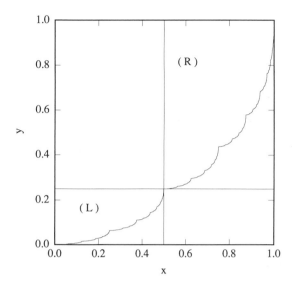

FIGURE 10.3: Two affine transformations of the coordinate system mapping the unit square into rectangles by the L and R transformations given in equation (10.10). A combination of the two transformations reproduces the original self-affine curve $y = M(x)$, calculated for $p = 0.25$.

These equations are same as the equations (6.7) for the *measure* for the multiplicative process discussed in section 6.2. An example of the probability of success under the strategy of bold play, $M(x)$, as a function of x with $p = 0.25$ is shown in figure 6.3 and also in figure 10.3. This probability (or measure) is an increasing function of x, and the function is singular in the sense that it almost everywhere has a slope of zero. The probability of success increases only on a set of points of Lebesgue measure zero where the left-hand derivative of $M(x)$ is $+\infty$ and the right-hand derivative is 0 (see Billingsley, 1983, for a clear discussion). The curve $M(x)$ has the length $L = 2$, and is therefore *not* a fractal curve — it is a fractal measure as discussed in section 6.2.

The equations (10.10) represent an invariance of the measure $M(x)$ under affine coordinate transformations. These transformations are given by

$$L: \quad (x, y) \quad \rightarrow \quad \left(\tfrac{1}{2}x, py\right),$$
$$R: \quad (x, y) \quad \rightarrow \quad \left(\tfrac{1}{2}, p\right) + \left(\tfrac{1}{2}, (1 - p)y\right). \tag{10.11}$$

These transformations map the curve $y = M(x)$ into itself — the curve is self-affine. There are *two* transformations. The transformation L maps the unit square into the rectangle marked (L) in figure 10.3. This transformation simply scales down the x-axis by a factor of $\tfrac{1}{2}$ and the y-axis by a factor p. The R-transformation also scales down the x-axis by a factor of $\frac{1}{2}$, but the y-axis is scaled down not by p but by a factor $(1 - p)$. The R-transformation in addition translates the scaled-down unit square so that it is mapped into the rectangle marked (R) in figure 10.3.

Mandelbrot (1985, 1986) has discussed many aspects of self-affine fractal curves that may be generated by families of affine transformations. He introduces a class of *recursively* self-affine fractal curves for which the relations given in the table in the previous section hold. He also discusses additional dimensions that may be used to characterize self-affine fractal curves.

Chapter 11

Wave-Height Statistics

Waves at sea are not only interesting to watch — they are of vital importance to marine activities. A large fraction of losses of lives and ships are due to the action of large waves in rough weather. Detailed observations of wave-height, period, and other characteristics have therefore been performed at many locations all over the world.

We will here discuss wave data from Tromsøflaket, made available to us by the Norwegian Institute of Meteorology. The wave data for the period 1980–1983 have been obtained using a buoy to record the water level. The wave data are collected by recording the wave-height 2048 times at 2-second intervals every 3 hours, i.e., for a period of about 17 minutes. The wave-height is defined as the difference between the highest and lowest water level between two 'zero up-crossings' of the water level. Here the zero level is the running average water-level.

This massive set of observations is first analyzed by conventional methods. For instance one calculates the *significant wave-height* h_s as the average of the largest $1/3$ of the waves recorded in the 17-minute observation interval. The maximum wave-height recorded in the interval, h_{max}, is about $1.8\,h_s$. We will use h_s in our discussion since this is a robust measure of the wave-height. The reason is that it is quite probable that the maximum wave-height occurs outside the observation interval, which only represents 9.4% of the total time. In fact we may estimate the maximum wave-height as $1.8\,h_s$. Other parameters characterizing the waves — such as the average period T_s for the highest $1/3$ of the waves, the average period T_z of the zero crossings and the energy spectrum for the 2048 readings — are also calculated.

Much interest is related to the prediction of the '100-year wave,' i.e., the question: How can we best estimate extreme wave-heights on the basis of a limited record of observed wave-heights? A large body of research

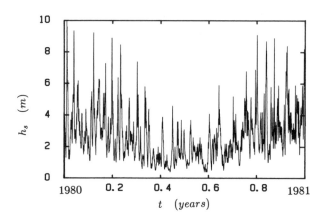

FIGURE 11.1: The maximum significant wave-height h_s at Tromsøflaket, observed every 3 hours in 1980 (The Norwegian Institute of Meteorology).

addresses such questions — see for instance Fjørtoft (1982) — and we shall not attempt to review this here. However, such predictions must be based on an understanding of the statistics of wave-heights — which may be *fractal*. We have therefore analyzed wave-height data taking the record $h_s(t)$ to be a fractal record similar to fractal Brownian motion.

11.1 R/S Analysis of the Observed h_s

In figure 11.1 we show the highest significant wave-height h_s for each day in 1980. The maximum h_s is 10.7 m and the highest recorded value of $h_{\max}$ is about 19 m. The seasonal variations are a significant feature of the wave-height data.

Interpreting the observed wave-heights as steps in a Brownian motion we make the identification $h_s(t) \rightarrow \xi(t)$, in terms of the notation introduced in previous chapters. With this identification we may naively just go ahead and calculate R/S as a function of the lag τ for the wave-heights on Tromsøflaket. As shown in figure 11.2 we find a very nice fit of the *Hurst law* $R(\tau)/S \sim \tau^H$ to the data with a Hurst exponent $H = 0.87 \pm 0.01$.

This rather high value of the Hurst exponent indicates that the wave-height statistics are *strongly non-Gaussian*. And since the observed value of H is significantly above $1/2$ we are tempted to conclude that the wave-height exhibits persistence.

A closer inspection of figure 11.2 shows, however, that there is some structure in the curve at lags τ near one year. This structure is clearly

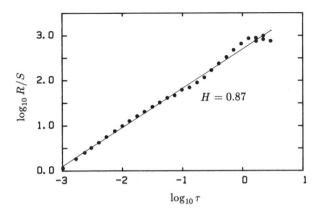

FIGURE 11.2: R/S as a function of the lag τ, for a process where the significant wave-height h_s for Tromsøflaket is considered to be the random process $\{\xi\}$. The line represents a fit of the Hurst law $R/S \sim \tau^H$ to the data with $H = 0.87 \pm 0.01$ (Frøyland et al., 1988).

related to the seasonal variations. The presence of periodic elements in the statistics is a complicating factor and there are no general rules for handling periodicities. Mandelbrot and Wallis (1969a) have considered the R/S statistics of sunspot activity, which have a well known 11-year cycle. Their results, shown in figure 11.3, give an apparent Hurst exponent of $H = 0.93$, which greatly exceeds $1/2$.

11.2 R/S for Seasonally Adjusted Data

There is little discussion in the literature on how to assess the influence of periodic elements on the estimated Hurst exponent (Mandelbrot and Wallis, 1969e). We have therefore adjusted the observations for seasonal effects. The result of this adjustment is that we may represent the observed data by a time series, as shown in figure 11.4.

The normalized wave-height has zero mean, $\langle \xi \rangle = 0$, and unit variance, $\langle \xi^2 \rangle = 1$, so that figure 11.4 may be directly compared to the fractal noise and the fractal functions in figures 9.4 and 9.5. It is clear from the record $X(t)$ in figure 11.4 that the excursions in X are large as compared to the 'noise.' The R/S analysis of the adjusted wave-height data shown in figure 11.5 now exhibits *two regions*. For times up to about 10 days, we find *persistent statistics* with $H = 0.92 \pm 0.02$. On the other hand, for lags above $\tau = 20$ days, we find $H = 0.52 \pm 0.02$, consistent with an *independent* process.

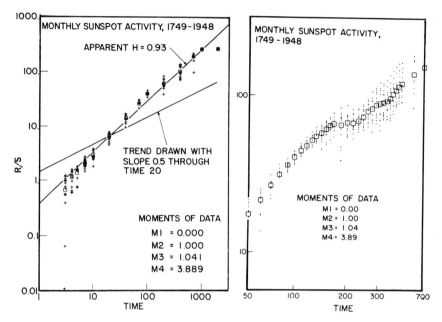

FIGURE 11.3: R/S as a function of the lag time τ for monthly Wolf numbers of sunspot activity. The 11-year cycle is clearly visible in the detailed portion shown in the right figure (Mandelbrot and Wallis, 1969a).

We still have a small anomaly near $\tau \sim 1$ year but it is much weaker. The distinctly persistent behavior for $\tau < 10$ days did not change from the R/S analysis of the original data shown in figure 11.2, which gives $H = 0.94 \pm 0.02$ for this range. This result indicates that the R/S analysis indeed gives a *robust* measure of the statistics of a time series, as discussed by Mandelbrot and Wallis (1969e).

As discussed in chapter 10, we may estimate the fractal dimension of a self-affine record such as $X(t)$ in figure 11.4 using the box-counting method — provided we remember that the result may depend on the size and shape of the initial box. In figure 11.6 we show the results of the box-counting method applied to the wave-height data.

With small boxes we measure the *local* fractal dimension. The smallest box that can be applied to these data corresponds to the resolution of the data which is $a = 0.1$ m and $\tau = 3$ hours. With this box size we find $D = 1.09 \pm 0.02$, which actually satisfies relation (10.4), $D = 2 - H$, with the Hurst exponent $H = 0.92$, obtained from the fit for $\tau < 10$ days. On the other hand, increasing the box size along the time axis to $\tau = 120$ hours gives $D = 1.52 \pm 0.03$, which is the value expected for an independent

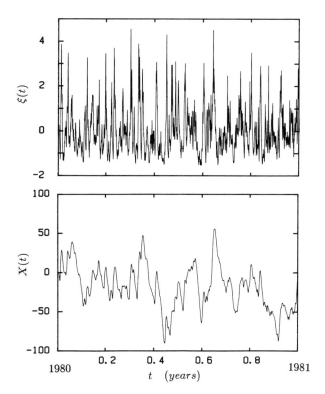

FIGURE 11.4: The normalized seasonally adjusted wave-height $\xi(t)$ as a function of time for the observations of h_s on Tromsøflaket. The bottom figure shows the cumulative sum $X(t) = \sum_{u=0}^{t} \xi(u)$ as a function of time (Frøyland et al., 1988).

random process. Again we find $D = 2 - H$, using the $H = 0.52$ obtained from the fit for $\tau > 20$ days.

It is important to remember that the crossover from $D = 1.09$ to $D = 1.52$ is *not* due to the crossover from the local to the global fractal dimension for self-affine records. The two fractal dimensions have both been determined using the highest wave-height resolution consistent with the data, i.e., 0.1 m, and both represent *local* dimensions. The observed crossover is due to a crossover from persistent to independent random behavior in local values of D.

We conclude that the wave-height statistics are persistent with a highly anomalous value of the Hurst exponent $H = 0.92 \pm 0.02$, as estimated both from the R/S analysis and from the determination of the box dimension of the fractal record. This persistence crosses over to an independent ran-

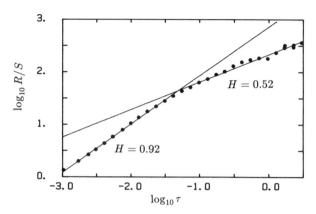

FIGURE 11.5: R/S as a function of the lag τ (in years), for a process where the seasonally adjusted and normalized significant wave-height ξ for Tromsøflaket is considered to be the step in a (fractional) random walk. The lines are fits of the Hurst law $R/S \sim \tau^H$. The fit for $\tau < 10$ days gives $H = 0.92 \pm 0.02$, and the fit for $\tau > 20$ days gives $H = 0.52 \pm 0.02$ (Frøyland et al., 1988).

dom process at a time scale of about 2 weeks. It should also be noted that this does *not* imply that the normalized wave-height ξ constitutes an independent *Gaussian* process in this range. It definitely does not, since it has a skewed distribution that does not allow for negative wave-heights. However, as discussed by Mandelbrot and Wallis (1969e), even extremely non-Gaussian independent processes controlled by either log-normal, hyperbolic or gamma distributions of the increments still give $H = 1/2$ in the R/S analysis, which is a very robust form of analysis.

Our observation of fractal statistics of the height of ocean waves must be taken into account in the prediction of extreme waves on the basis of observed wave-heights. However, to our knowledge fractal statistics that may be used for such predictions have not yet been developed. The full statistical consequences of the fractal nature of wave-height data remain to be explored.

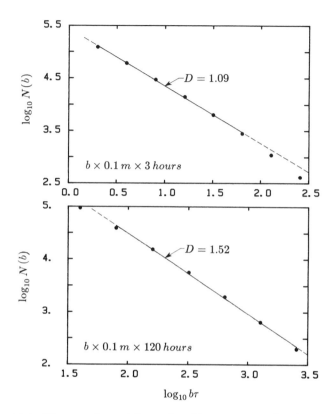

FIGURE 11.6: The number of 'boxes' $N(b)$ of size $b \times (a, \tau)$ as a function of the box scale factor b for the cumulative normalized wave-height $X(t)$, representing one data point every 3 hours, recorded at Tromsøflaket in the period 1980–1983. The top figure uses a minimum box of $a = 0.1$ m and $\tau = 3$ hours. The curve is a fit of $N \sim b^{-D}$ with $D = 1.09 \pm 0.02$. The bottom figure also uses $a = 0.1$ m but a time width $\tau = 120$ hours. The fit gives $D = 1.52 \pm 0.03$ (Frøyland et al., 1988).

Chapter 12

The Perimeter-Area Relation

For circles, squares, equilateral triangles and other polygons the ratio between the perimeter and the square root of the enclosed area,

$$\rho = (\text{Perimeter})/(\text{Area})^{\frac{1}{2}},$$

is independent of the size of the polygon. The ratio ρ is the same for all closed curves of the same *shape*. The ratio $\rho = 2\sqrt{\pi}, 4$, and $6/3^{\frac{1}{4}}$ for circles, squares and equilateral triangles, respectively.

For a collection of *similar* islands of different sizes, the ratio between the length of a nonfractal coastline of any island and the square root of its area is independent of the size of the island. However, for islands with fractal coastlines the length $L(\delta)$ of the coastline depends on the yardstick δ used to measure its length, and $L(\delta) \to \infty$, as $\delta \to 0$. On the other hand, the area $A(\delta)$ of the island, measured by covering it with squares of area δ^2, remains finite as $\delta \to 0$. Mandelbrot shows that for fractal curves the divergent ratio ρ should be replaced by the modified ratio given by

$$\rho_D = (\text{Perimeter})^{1/D}/(\text{Area})^{\frac{1}{2}} = [L(\delta)]^{1/D}[A(\delta)]^{-\frac{1}{2}}, \qquad (12.1)$$

for each of the islands. Here D is the fractal dimension of the coastlines of the similarly shaped islands. The ratio ρ_D is independent of the size of the island — but it does depend on the yardstick chosen.

The perimeter-area relation as expressed by equation (12.1) follows from the definition of the fractal dimension D contained in equations (2.3) and (2.4). This is seen by comparing two similar islands of different area as shown in figure 12.1. The area and the length of the coastline of each of the islands is measured by using an *area-dependent* yardstick $\delta_i^* = \lambda\sqrt{A_i(\delta)}$ for the i-th island. The area of the i-th island, when measured with a fixed yardstick δ, is $A_i(\delta)$, and the parameter λ is some arbitrary small number

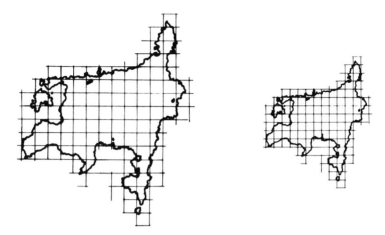

FIGURE 12.1: Two similar islands measured with area-dependent yard-sticks.

— say 0.0001. The length of the coastline of the i-th island equals the perimeter of the polygon with N_λ segments of length δ^* and $L_i(\delta^*) = N_\lambda \delta_i^*$, to this approximation. Now the important observation is that for similarly shaped islands N_λ is *independent* of the size of the island. However, from equation (2.3) it follows that the length of the coastline of the i-th island is $L(\delta) = L_i^0 \delta^{(1-D)} = L_i(\delta^*)(\delta/\delta^*)^{(1-D)}$. We therefore obtain the expression

$$L_i(\delta) = N_\lambda \delta^{(1-D)} \delta^{*D}.$$

Now we express δ^* in terms of $A(\delta)$:

$$L_i(\delta) = N_\lambda \lambda^D \delta^{(1-D)} \sqrt{A_i(\delta)}^D,$$

and it follows that the ratio

$$\rho_D(\delta) = \frac{[L_i(\delta)]^{1/D}}{[A_i(\delta)]^{\frac{1}{2}}}$$

is independent of the size of the island. However, the ratio $\rho_D(\delta) = N_\lambda^{1/D} \lambda \delta^{(1-D)/D}$ does depend on the given yardstick length δ, and on the arbitrary factor λ used. Therefore, although $\rho_D(\delta)$ is related to the *shape* of the islands, it also contains arbitrary factors and so far we have no general measure of shape. We conclude that islands similar in form satisfy the following *perimeter-area relation* due to Mandelbrot

$$L(\delta) = C \delta^{(1-D)} \sqrt{A(\delta)}^D . \tag{12.2}$$

This holds for any given yardstick δ small enough to measure the smallest island accurately. The constant of proportionality $C = N_\lambda \lambda^D$ depends on the arbitrary parameter λ. The relation (12.2) is the basis for the practical determination of the fractal dimension in several interesting cases discussed in the following sections.

As an example of the perimeter-area relation consider the quadratic Koch island in figure 2.9. The initiator is a square with sides a. The generator adds a small 'peninsula' and subtracts an equally large 'bay' as each edge of the previous generation is replaced. Therefore the iteration process does not change the area $A(\delta) = a^2$. The n-th generation perimeter $L_n = 4 \cdot 8^n \left(\frac{1}{4}\right)^n \cdot a$ is the length of the coastline when measured with a yardstick of length $\delta = \left(\frac{1}{4}\right)^n a = \left(\frac{1}{4}\right)^n \sqrt{A}$. Therefore we may write the generation number as $n = \ln\left(\delta/\sqrt{A}\right)/\ln\frac{1}{4}$, and we find $L_\delta = 4 \cdot 8^n \delta = 4\delta^{1-D} A^{D/2}$. Here $D = 3/2$ is the fractal dimension of the coastline of the quadratic Koch island. We conclude that quadratic Koch islands satisfy the *perimeter-area* relation in the form of equation (12.2), with the constant of proportionality given by $C = 4$.

Another example is provided by the triadic Koch curve. The island is considered to be the area between the original initiator, i.e., a straight line of length a, and the limiting Koch curve (see figure 2.8). The enclosed area is $A = \sqrt{3}a^2/20$, and in the n-th generation the total length of the coastline is $L_n = a + 4^n \delta$, with the yardstick length $\delta = \left(\frac{1}{3}\right)^n a$. This gives the relation

$$L_\delta = a + a^D \delta^{(1-D)}.$$

We see that if we neglect the first term on the right-hand side we recover the *perimeter-area relation* (12.2). The neglected term represents the non-fractal straight line basis of the islands. This example illustrates that we may expect the perimeter-area relation to hold only in the limit of small yardsticks δ, where the length of the fractal coastline dominates any *regular* part of the coastline. More complicated examples in which various regions of the coast have different fractal dimensions are easily constructed and one concludes that eventually the perimeter-area ratio ρ_D is dominated by the region with the highest value of the fractal dimension.

12.1   The Fractal Dimension of Clouds

Lovejoy (1982) investigated the geometry of cloud and rain areas in the very large size range from 1 km^2 to $1.2 \cdot 10^6$ km^2, and found that the cloud perimeter $P = L$ is related to the cloud area A by the perimeter-area relation (12.2), with the perimeter fractal dimension $D = 1.35 \pm 0.05$. His

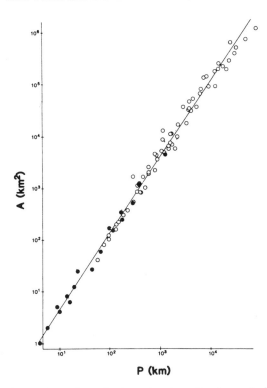

FIGURE 12.2: Area plotted against perimeter of rain and cloud areas. •, Radar rain areas; ∘, satellite cloud areas (Lovejoy, 1982).

results are reproduced in figure 12.2. Rain areas were studied by digitizing radar pictures at a resolution of 1 × 1 km. The radar senses the reflected microwave radiation primarily from the large rain drops. A connected set of pixels for which the rain rate exceeded 0.2 mm/hour (which corresponds to a light drizzle) constituted a rain area. The area A of a rain area is the sum of a connected set of such pixels, whereas the perimeter P or L is given by the number of rain pixels that have no-rain pixel neighbors. The curvature of the earth limits the radar analysis to rain areas $\leq 40,000$ km^2. Lovejoy's results for the rain areas are represented by the black dots in figure 12.2.

Infrared pictures of Indian Ocean clouds were sampled on a grid of 4.8 × 4.8 km. The pictures were obtained by the geostationary operational environment satellite (GOES). In order to avoid effects of varying picture element size only data in the relatively undistorted region of 20°N and 20°S, and within ± 30° of longitude of the subsatellite point were used. The GOES infrared sensor responds primarily to the blackbody radiation emitted by the clouds and the surface. For clouds more than about 300 m thick, the infrared channel yields an estimate of the cloud-top temperature. Physically, the rain and cloud fields are closely related in the tropics, since they both occur in regions of convective draft, which causes moist warm surface air to rise, cool by adiabatic expansion and form clouds and rain in the resulting condensation processes. Much of these clouds is supercooled water rather than ice. Pixels with a temperature below $-10°$ C were considered to be cloud pixels. The cloud area delineated by the $-10°$C threshold thus contains both cumulus and cirrus clouds. As before the cloud area was determined by counting the number of cloud pixels in the connected set defining a cloud, whereas the perimeter length was obtained by counting the number of cloud pixels that had noncloud neighbors. The results for many different cloud areas are plotted as open circles in figure 12.2. However, it should be noted that for the satellite pictures perimeters have been increased by the factor 1 km/4.8 km$)^{(1-D)}$, to compensate for the difference in resolution of the radar and satellite data. This factor accounts for the prefactor $\delta^{(1-D)}$ in equation (12.2).

The most remarkable aspect of the results in figure 12.2 is the absence of any apparent bend or kink over a range of six orders of magnitude in area. The results in figure 12.2 represent clouds of different macroscopic shape, yet they still are all points on the same line in the area-perimeter plot. Selecting a different temperature in the cloud definition changes their area and perimeter, but in such a way that the points representing the clouds only shift position on the same line. This area-perimeter relation determines the fractal dimension of the cloud perimeters to be $D = 1.35 \pm 0.05$. It is important to realize that this result leads to the conclusion that there is no characteristic length scale ξ, in the range 1 km to 10,000 km, for the processes involved in clouds. This is a remarkable result since one might have guessed at $\xi \sim 10$ km — the height of the atmosphere. In addition clear air Doppler radar wind measurements show no length scales in the range 4 to 400 km, and Doppler wind spectra in rainy regions exhibit no length scale in the range 1.6 to 25 km. The conclusion is that the atmosphere exhibits no intrinsic length scale and is best described as being *fractal* — in fact it seems that clouds are *self-affine fractals*.

Rys and Waldvogel (1986) studied the fractal shape of hail clouds. These severe convective clouds were observed using radar and they found the perimeter-area relation shown in figure 12.3. Each point in the figure corresponds to a particular time (at 1-minute intervals) during the evolution of the hail storm.

The figure clearly shows that there is a crossover at a perimeter P_0 marked by the dotted line in figure 12.3. This corresponds to a crossover diameter $d_0 \simeq P_0/\pi = 3 \pm 1$ km. The results of linear fits of the perimeter-

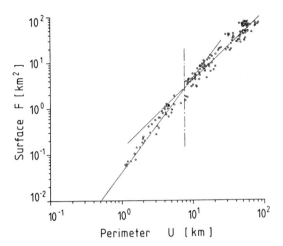

FIGURE 12.3: Log-log plot of the perimeter-area relation for hail storms. Every point corresponds to a particular time during the temporal evolution of a storm. From the linear fits (shown as full lines) of 24 different hail storms the averaged fractal dimensions in equation (12.3) were obtained (Rys and Waldvogel, 1986).

area to the observed values are

$$D = 1.36 \pm 0.1 , \quad \text{for} \ \ P > P_0 ,$$
$$D = 1.0 \pm 0.1 , \quad \text{for} \ \ P < P_0 . \tag{12.3}$$

Above the crossover these results give the same fractal dimension for the cloud perimeter as the one observed by Lovejoy. By contrast, for length scales below approximately 3 km, the authors find that the severely convective hail storms have perimeters that are not fractal.

Theory of the Fractal Dimension of Clouds

Hentschel and Procaccia (1984) calculate the fractal dimension of cloud perimeters to be in the range $1.37 < D < 1.41$ on the basis of their theory of relative turbulent diffusion. The structure of their theory is interesting. The central question is: How can the cloud change its overall *shape* in time and still exhibit a *universal* [1] fractal structure? Clearly the time development of a cloud is not universal. The answer according to Hentschel and Procaccia lies in a subtle interplay between length scales and time scales

---

[1] Here the word *universal* is used to indicate that the fractal structure is independent of initial conditions such as the size, height or other parameters specifying the initial cloud.

in fully developed homogeneous turbulence. In fact they have previously shown that their model of fractally homogeneous turbulence (Hentschel and Procaccia, 1983a) leads to a fractal dimension D_μ of the turbulent field in the range $2.50 < D_\mu < 2.75$.

Hentschel and Procaccia propose a simple model of a cloud. The state of the atmosphere is given by specifying such parameters as the temperature, pressure, water content and droplet sizes, in addition to the turbulent velocity field. In Lovejoy's experiments either droplet radius or temperature was used as a criterion for deciding whether a given part of the atmosphere was part of a cloud or not. In the proposed model one chooses for instance the local temperature θ as a variable that is transported by the turbulent field – but it is assumed that the turbulent field is not affected by it. An equally acceptable choice would be the cloud droplet radii. The state of the atmosphere is given by specifying the function $\theta(x, y, z)$, and therefore the model ignores the cloud physics relating such factors as temperature, humidity and droplet sizes and considers only one so-called *passive scalar* variable — here θ — and the turbulent wind field transporting this variable. For a given point $\mathbf{r} = (x, y, z)$ in space the function $\theta(x, y, z)$ has some value and therefore specifies a point (θ, x, y, z) in an $E = 4$-dimensional space. The set of points $\mathcal{S} = \{(\theta, x, y, x)\}$ in four-dimensional space is *fractal*, with dimension $D_\theta = 4 - H$, where the codimension is H. Defining a cloud by the condition $\theta \leq \theta_0$ defines a region in the four-dimensional space describing the atmosphere, which also has the fractal dimension D_θ. However, the surface of the cloud defined by the condition $\theta(x, y, z) = \theta_0$ is a set of points $\mathcal{C}$ in the ordinary $E = 3$-dimensional space. This set of points is the intersection of the set of points $\mathcal{S}$ and the set of points $\mathcal{S}_0 = \{(\theta_0, x, y, z)\}$, which has the fractal dimension $D_0 = 3$. Formally we have $\mathcal{C} = \mathcal{S} \cap \mathcal{S}_0$. The fractal dimension of the cloud surface $\mathcal{C}$, is $D = 3 - H$. This follows from the 'rule of thumb' given by Mandelbrot (1982, p. 365) stating that the codimension of the intersection of two independent sets in E-space almost surely[2] equals the sum of the codimensions of the intersecting sets. In the present case this gives

$$(4 - D) = (4 - D_\theta) + (4 - D_0)$$

therefore $(4 - D) = H + 1$, or $D = 3 - H$ as stated. By intersecting the cloud surface in three-dimensional space by a plane parallel to the ground, we obtain a set of points $\mathcal{P}$ — the perimeter, which, using Mandelbrot's rule of thumb, must have a fractal dimension $D_P = 2 - H$. We conclude

---

[2] A simple counter-example is provided by the intersection of the baseline (the original initiator) and the triadic Koch curve. This intersection is a Cantor set with the fractal dimension $D = \ln 2/\ln 3 = 0.63\ldots$, whereas the rule of thumb gives $(2 - D_\cap) = (2 - \ln 4/\ln 3) + (2 - 1)$, or $D_\cap = (\ln 4/\ln 3) - 1 = 0.26\ldots$.

FIGURE 12.4: Clouds simulated by the fractal sums of pulses model. Areas that have a rain rate below a given threshold are black. The rain rate is on a log scale with the highest rate being white. The exponent $H = 0.6$ was used in the simulation (Lovejoy and Mandelbrot, 1985)

that if we obtain an estimate for the codimension H, we obtain estimates for the fractal dimensions of clouds. From Lovejoy's data we conclude that for clouds and rain areas $H = 0.65 \pm 0.05$.

Hentschel and Procaccia (1984) estimate the fractal dimension of the cloud perimeter to be in the range

$$1.37 < D_P < 1.41,$$

in excellent agreement with the results reported by Lovejoy (1982). We find it remarkable that such a simplified model of the atmosphere is capable of calculating the fractal dimension of clouds in agreement with observation. Maybe in the future other examples of observed fractal dimensions may be calculated from physical models of the system.

Note that the cloud in four-dimensional space is self-affine not self-similar. However, the surface of the clouds in three-dimensional space may be self-similar. This distinction has been emphasized by Voss (1985a,b), who has generated truly remarkable pictures of fractal clouds — with a visual quality that had previously only been achieved by painters.

Lovejoy and Mandelbrot (1985) propose a self-similar fractal model for rain fields. In this model the rain field is considered to be the superposition of 'pulses.' Each pulse covers an area A, and an increment in the rain rate

$\Delta R = \pm A^H$. The rain rate is chosen to randomly increase or decrease the total rain rate over the area A. The area is assumed to have a *hyperbolic* probability distribution: $\Pr(A > a) \sim a^{-1}$. Figure 12.4 shows the result obtained with $H = 0.6$. The resulting pictures look quite realistic. We have also generated cloud pictures by another algorithm, discussed in chapter 13.

Lovejoy and Schertzer (1985) point out that the atmosphere is layered and clouds are not self-similar but self-affine, the vertical direction being singled out from the horizontal direction. In addition the effects of the Coriolis force lead to rotations. They also introduce a modification of the fractal sum of pulses model in which the pulses are distributed not uniformly but on a fractal set of points in space. Again realistic-looking pictures of clouds are generated. Mandelbrot (1986) comments that the self-affine function $\theta(x, y, z)$ transforms in a self-similar manner in the horizontal directions x, y, when $(x, y) \to (rx, ry)$, and scales differently in the z and θ directions using r^G and r^{GH}, respectively. Here, G is an extra Lipschitz-Hölder exponent. In view of our discussion of multifractal measures it appears likely that one should in fact consider clouds to be fractal measures $M(\mathbf{r})$, defined in space. The complete characterization of this measure probably involves a spectrum of fractal dimensions. In short clouds are multifractals, a view proposed by Lovejoy and Schertzer (1985) (see also Lovejoy et al., 1987).

12.2   The Fractal Dimension of Rivers

Mandelbrot (1982) has pointed out that rivers satisfy the length-area relation in equation (12.2). We discuss here some of the results presented by Hack (1957), but earlier and later work on this subject exists. Hack studied rivers in Virginia and Maryland. In figure 12.5 the drainage area A above a particular locality in a river system is indicated. In the same figure the length L is measured along the *longest* stream to the drainage divide. Figure 12.6 shows the results obtained by Hack. The areas were determined from topographic maps and in a few cases from aerial photographs, by use of a planimeter. The length of streams was generally determined with a map measure, on maps or aerial photographs, along the stream and following meanders and bends.

The straight line in figure 12.6 describes all the data well and is given by

$$L = 1.4\sqrt{A}^{D}, \text{ with } D = 1.2 . \tag{12.4}$$

Hack has also checked this relation by plotting 400 similar measurements made by Langbein et al. (1947) in the northeastern United States, and he finds the same relation (12.4). Thus it is fairly well established that in

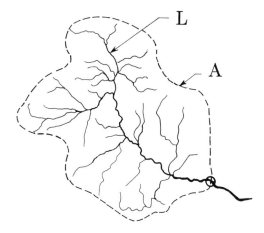

FIGURE 12.5: Drainage area A and the length L of the longest river above a given location.

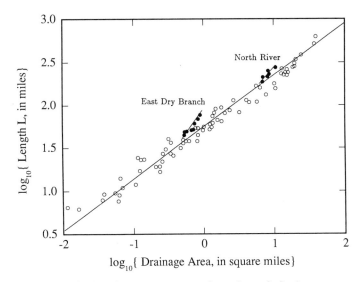

FIGURE 12.6: Relation between stream length and drainage area (Hack, 1957).

the northeastern United States the length of a stream at any locality is, on the average, proportional to $\frac{1}{2}D = 0.6$ power of its drainage area at the locality — regardless of the geological or structural characteristics of the area. However, Hack has also considered two regions in the western United States and found that there $D = 1.4$. Therefore the relation (12.4) is not entirely general.

The solid circles in figure 12.6 represent two specific cases of departure from the general relation between length and area. The coefficient in the length-area relation is 1.4 (when units of miles and square miles are used), but is found to vary in the range 1 to 2.5, with an average value of 2.0 in sandstone.

Following Mandelbrot it is tempting to conclude that the result (12.4) gives the fractal dimension of rivers and streams to be $D = 1.2$. However, river systems are not self-similar, and the arguments leading to the length-area relation (12.2) therefore do not apply. The lack of self-similarity was noted by Hack and expressed as follows:

> *It is obvious that on any stream the length must increase downstream more rapidly than the 0.6 power of the area between all the major tributaries. There are stream reaches along which no large tributaries enter for long distances, and in these reaches the length must increase at a more rapid rate; this general principle applies to short as well as to long streams.*

Note that for rivers with a constant drainage width d, the length-area relation would become $L = A/d$, and one would conclude that $D = 2$ for such rivers. On the other hand, if the drainage basin retained the same shape as it enlarged we would find $D = 1$.

Hack presents a model for the river system based on an analysis of the stucture of river systems introduced by Horton (1945). Horton labels rivers and tributaries by assigning each stream an *order* in such a way that streams that have no tributaries have stream order $i = 0$. Streams of order i have tributaries of order $(i-1), (i-2), \ldots$. All the streams in a drainage area may be labeled in this way, with the result that the principal stream is assigned the highest order s. The length ratio r_l, which is the ratio of the average length of streams of one order to the average length of streams of the next lowest order, is approximately a constant and independent of stream order. The average length of a stream of order i is therefore given by

$$L_i = r_l L_{i-1} = r_l^i L_0 \; ,$$

where L_0 is the average length of the smallest streams. The bifurcation ratio r_b is the ratio between the number of streams of one order and the number of streams of the next higher order:

$$N_i = r_b N_{i+1} = r_b^{s-i} N_s \; .$$

The bifurcation ratio is approximately independent of stream order. Note that $N_s = 1$, since there is only one stream of the highest order.

Since the '*the length of overland flow is about the same for all streams*' (Hack, 1957) the area that drains directly into a second-order stream must have the same average width, d, as the area that drains directly into the first-order stream, and as a consequence this term increases in proportion to the length. Hack concludes that the drainage area of order s is given by

$$A = \sum_{i=0}^{s} L_i \, dN_i = A_0 r_b^s \sum_{i=0}^{s} (r_l/r_b)^i \; .$$

Clearly the overland drainage length introduces a length scale into the problem which prevents a scaling behavior. Ignoring this term we find that the order s may be written $s = \ln(A/A_0)/\ln r_b$, and we find for a basin of order s

$$L = L_0 \sqrt{A/A_0}^{D_s}, \text{ with } D_s = 2 \ln r_l/\ln r_b \; .$$

Hack discusses an example in which $r_l = 2.4$ and $r_b = 3.2$, i.e., $D_s = 1.5$. When the overland drainage term is included the length-area relation is no longer a power law. Approximating it by one leads Hack to the value $D_s = 1.30$, compared with the observed $D = 1.22$ for this area.

The essence of Hack's model relating the bifurcation ratio r_b and the length ratios r_l is intuitively relevant to the observed power-law relation between length and area. However, how to define a fractal dimension for river systems remains obscure, and clearly the geometry of streams and rivers requires further study.

Chapter 13

Fractal Surfaces

Much of the broad interest in fractals has probably been generated by the striking computer-generated pictures of landscapes. The beautiful color pictures in Mandelbrot's last book (Mandelbrot, 1982) of valleys and of a rising 'earth' seen from a 'moon' have captured the imagination of scientists and laymen alike. An early popular discussion by Gardner (1976) of Mandelbrot's work probably helped much in building interest — at least it was my first contact with these ideas. Even more spectacular pictures with haze in the valleys, also generated by R. F. Voss, have been published in a popular account in a widely circulated microcomputer journal (Sørensen, 1984). Indeed the irregularity of the earth's topography over a large range of length scales indicates that useful models of landscapes may be obtained using fractals.

In this chapter we first discuss simple fractal surfaces. Then we discuss our experience with the generation of landscapes. In the next chapter we discuss the recent experimental evidence for the fractal structure of surfaces.

13.1 The Fractal Koch Surface

A simple way of constructing a manifestly fractal surface is to slide the triadic Koch curve along a direction perpendicular to the plane of the Koch curve for a distance ℓ, as shown in figure 13.1. In order to measure the area of such a surface, we would try to cover it with strips of length ℓ and width δ. With small yardstick areas $a = \delta \times \delta$ we would need

$$N(\delta) = \frac{\ell}{\delta} \times \frac{1}{\delta^{D_1}}$$

212

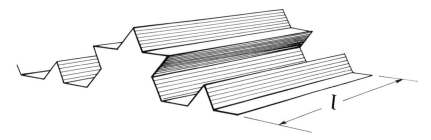

FIGURE 13.1: A triadic Koch surface. $D = 2.262\ldots$.

such pieces to cover the surface. Here the fractal dimension of the Koch curve is given by $D_1 = \ln 4/\ln 3$ as before. The first term in the expression for $N(\delta)$ simply states that the number of a's needed per strip is ℓ/δ, whereas the second term is the number of segments of length δ needed to cover the Koch curve — see equation (2.6). The measure function M_d is given by

$$M_d = N(\delta)\delta^2 = \ell\delta^{d-D_1-1}.$$

Since this measure remains finite only for $d = D$ given by

$$D = 1 + D_1 = 2.262\ldots,$$

it follows from equation (2.3) that D is the Hausdorff-Besicovitch dimension of the surface.

This result is another example of one of Mandelbrot's 'rules of thumb' for fractal sets (Mandelbrot, 1982, p. 365): For a set $\mathcal{S}$ that is the product of two independent fractal sets $\mathcal{S}_1$ and $\mathcal{S}_2$, the fractal dimension of $\mathcal{S}$ equals the sum of the fractal dimensions of $\mathcal{S}_1$ and $\mathcal{S}_2$.

Here we use the set of points $\mathcal{S}_1$ defined by the triadic Koch curve in $E = 2$-dimensional space having dimension D, and we have generated a set $\mathcal{S}$ by the multiplication of the set $\mathcal{S}_2$, which is a line segment of length ℓ in $E = 1$-dimensional space. Thus, from any point in $\mathcal{S}_1$ we have generated a line of points in $E = 3$-dimensional space. The resulting set has dimension $D = D_1 + D_2$, with $D_2 = 1$ for the line.

Fractal surfaces of any dimension in the range $2 \leq D \leq 3$ may be generated in a similar manner — but they are not realistic models of landscapes and other irregular surfaces, which in general are much more isotropic in the horizontal plane than the translated Koch curve surfaces. Perhaps they could be used for surfaces resulting from directional wear, abrasion or polishing.

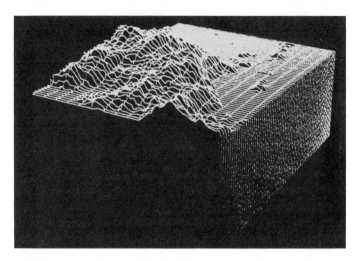

FIGURE 13.2: The first example of fractal Brownian islands. $D = 2.5$ (Mandelbrot, 1975b).

13.2   Random Translation Surfaces

A simple way of generating a more reasonable surface is to add to the height $z(x,y)$ generated by the translation of a fractal curve successive layers of similar profiles obtained by rotating the first surface. Let $z_D(x,y)$ be the surface profile generated by sliding a fractal curve of dimension D in the x, z-plane along the y-axis. We rotate this surface an angle ϕ in the x, y-plane to generate the surface $z_D(x,y \mid \phi)$, and finally we multiply the height by a factor h to define the surface $z_D(x,y \mid h,\phi) = h\, z_D(x,y \mid \phi)$. Using such surface profiles we may generate fractal surfaces of height $Z(x,y)$ given by

$$Z(x,y) = \sum z_D(x,y \mid h,\phi). \tag{13.1}$$

If we add only a few layers of fixed h and random ϕ, we readily generate interesting surfaces which — at least to the resolution of our computations — have the fractal dimension of the generating fractal surface or $D(Z) = D(z)$. The question — what is the fractal dimension $D(Z)$, in general? — is probably a very difficult one.

As an example of the complexities involved in the definition of the fractal dimension of such surfaces consider one of the early models proposed by Mandelbrot (1975b). The simplest model is one in which the function z_D is simply a step function, corresponding to a horizontal plateau broken along the straight line $x = 0$ in the x, y-plane with unit difference in height along the resulting fault. This generating surface is not fractal and

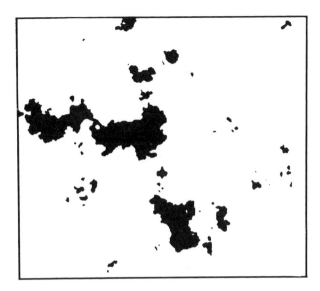

FIGURE 13.3: The first example of fractal Brownian coastlines. $D = 1.5$ (Mandelbrot, 1975b).

we have $D(z) = 1$. To generate $Z(x, y)$ choose ϕ with uniform probability in the range 0 to 2π, and choose $h = 1/\sqrt{n}$, for the n-th stage of construction (thus making an individual cliff negligible in size, compared with the cumulative sum of the other cliffs). The resulting surface, shown in figure 13.2, is *fractal* with dimension $D = 5/2$, in spite of the fact that the generating surface is *not* fractal. Mandelbrot calls such surfaces generated by an infinite number of layers *Brownian surfaces* since any vertical section through the surface generates a curve characteristic of Brownian motion. The surface satisfies the scaling relation

$$Z(\lambda x, \lambda y) = \lambda^H Z(x, y)$$

in distribution for any value of λ, and the codimension $H = 3 - D = 1/2$.

This scaling relation shows that the fractal surface is self-affine, not self-similar. It is a generalization to higher dimensions of the self-affine fractals discussed in chapter 10. Note that we must distinguish local and global fractal dimensions for self-affine surfaces, just as for self-affine records. The question of how to deal properly with fractal surfaces in practice has not yet been completely resolved. For recent discussions see Mandelbrot (1985) and Voss (1985a,b).

By filling the landscape with 'water' to a given level, coastlines and islands appear, as shown in figure 13.3. The fractal dimension of the coastlines obtained by the intersection of a plane with the surface is $D = 2 - H = 3/2$, well above the value of about 1.3 observed for various coastlines.

Mandelbrot (1975b) in addition states that the islands satisfy the Korčak distribution

$$\text{Nr}(A > a) \sim a^{-B}, \tag{13.2}$$

where $\text{Nr}(A > a)$ is the number of islands having an area A greater than a prescribed value a. This relation is also called the *number-area relation*. The exponent B is given by $B = \frac{1}{2} D_{\text{coast}} = 3/4$. The observed value for the earth is $B \sim 0.65$, according to Mandelbrot, and ranges from 0.5 for Africa (one enormous island and others whose size decreases rapidly) up to 0.75 for Indonesia and North America (less overwhelming predominance of the largest islands).

13.3   Generating Fractal Surfaces

We have made a number of simulations of surfaces that look like natural landscapes, at least to the extent permitted by the simple plotter graphics available to most. It has been an interesting experience, well worth pursuing before more fancy graphics are attempted.

From the discussion in the previous section it is clear that by superimposing an infinite number of randomly oriented nonfractal surfaces a fractal surface may indeed be obtained. For practical purposes it is better to superimpose only a few *fractal* translation surfaces with random orientations. If in addition the generating translation surfaces are random, we expect to get reasonable surfaces with a rather limited amount of computational effort.

When we look at a landscape we only see a certain area — this defines an upper length scale $L_{\max}$. In addition we have a minimum length scale $L_{\min}$ determined by the resolution of the eye or the photographic film, or here by the numeric precision chosen for the x, y-coordinates. If we were using sinusoidal translation surfaces to generate surface relief, we could easily produce a rolling landscape. With a spatial frequency f we get a translation surface

$$z(x, y) = C_f \sin(f\, x)\,,$$

with an amplitude C_f. Now the lowest frequency of interest is given by $f_{\min} \sim 1/L_{\max}$, since lower spatial frequencies simply appear to be constant additions to the height all over the surface. The highest spatial frequency of interest is given by the resolution $f_{\max} \sim 1/L_{\min}$, since finer detail will not be resolved.

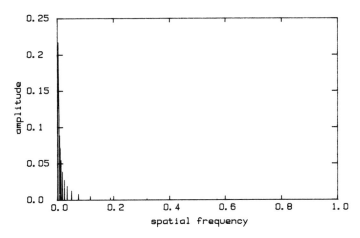

FIGURE 13.4: The amplitude spectrum for $f_0 = 0.004$, $\alpha = 1.1$ and $\beta = 1.1$.

We have chosen to generate translation surfaces from superpositions of oscillating functions. We first define the *frequency spectrum*, by specifying the spatial frequencies in the form

$$f_j = f_{j-1}^{\alpha} = f_0^{\alpha^j} , \quad \text{for } j = 1, 2, \dots . \tag{13.3}$$

This spectrum is *discrete* and is defined by the exponent α and the lowest frequency f_0. The base frequency, f_0, must be larger than f_{min} to produce visible effects. Increasing f_0 will result in pictures in which more of the landscape is seen in the observation window. The parameter α controls to a large degree the general appearance or type of landscape generated. We have chosen $0.7 \leq \alpha \leq 1.4$. Low values tend to generate alpine landscapes, whereas high values of α tend to generate rather smooth landscapes, like those common to southern Norway. The iterative form of the frequency spectrum (13.3) was chosen because it is a scaling form.

In general, landscapes have the highest amplitudes at low frequencies, and we have chosen a self-similar or scaling form for the *Fourier amplitudes*:

$$C_f \sim f^{-\beta} . \tag{13.4}$$

The constant of proportionality in this equation is irrelevant since the final picture later will be scaled to fit inside a specified format. The parameter β should be chosen in the range 1.05–1.4 in order to obtain landscapes that look 'natural.' The lowest values are suitable for alpine landscapes. This form of the amplitude spectrum tends to generate self-similar curves. In figure 13.4 we have shown an amplitude spectrum with the spatial frequency scaled so that the maximum frequency considered is 1.

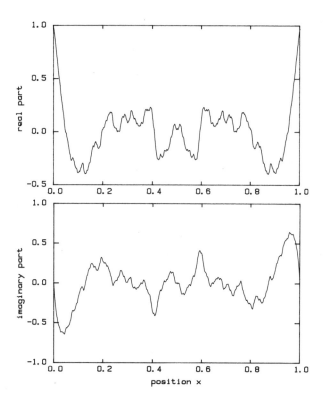

FIGURE 13.5: The real and imaginary parts of $\mathcal{Z}$ as functions of position x for $f_0 = 0.004$, $\alpha = 1.1$ and $\beta = 1.1$.

In order to obtain height as a function of distance we perform a complex Fourier-transform:

$$\mathcal{Z}(x) = \sum_j C_{f_j} \exp(2\pi f_j x) \ .$$

We have used the standard *fast Fourier-transform* algorithm to evaluate $\mathcal{Z}$. The resulting surface profile has real and imaginary parts: $z' = \Re\mathcal{Z}$ and $z'' = \Im\mathcal{Z}$, as shown in figure 13.5. The irregular fluctuations of z' and z'' are caused by the discrete nature of the frequency spectrum, which has irrationally related components.

We generate real translation surfaces from the profile $\mathcal{Z}$ by forming the expression

$$z(x, y \mid \psi) = z'\cos\psi + z''\sin\psi,$$

where ψ is an angle chosen at random in the range 0–2π. The resulting curve $z(x, 0 \mid \psi = 0.616)$ is shown in figure 13.6.

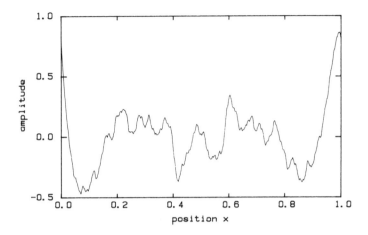

FIGURE 13.6: The profile $z(x, 0 \mid \psi = 0.616)$.

By the addition of a few such randomly oriented layers we generate a landscape. In order to enhance the structure of the landscape we fill water to some level and view it from a height we choose. Figure 13.7 results from the superposition of 12 layers of height profiles $z(x, y \mid \psi, \phi)$, with random phase angles ψ and rotated a random angle ϕ. For figure 13.7 we used a resolution of 1024 points, corresponding to unit distance in both the x- and y-directions. However, we have displayed only a central portion of 64×64 grid points in the x, y-plane. The picture is generated by drawing 64 height profiles of the form $\mathcal{Z}(x, y_i)$ with $i = 1, 2, \ldots, 64$, and in such a way that the pen is lifted if the next point is hidden by a contour already drawn.

The resolution in the previous figure is clearly too crude, and in the following pictures we have used a grid of 256×256 points in the x, y-plane, taken from the middle portion of the 1024-point Fourier-transformed z-curves. We find that 30 height profiles perpendicular to the line of sight, with sections in front hiding later sections, give reasonable results. In the regions filled with water we draw all of the 256 profile lines.

In order to get acceptable landscapes it is necessary to add perspective. A simulation of the curvature of the earth is needed in coastal landscapes in order to avoid the impression that the landscape is cut off at the horizon. An example using these features after the superposition of four layers is shown in figure 13.8. The picture might well have been taken in the region of granite rocks in southern Norway. The smooth appearance is caused by the rather large value of $\alpha = 1.4$, which causes the frequency spectrum to space out rapidly with increasing spatial frequency f. Decreasing the α to

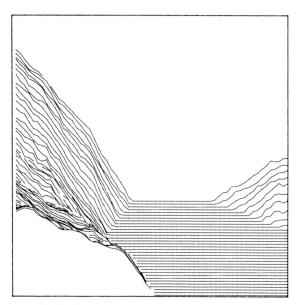

Figure 13.7: A landscape generated by 12 layers. $\alpha = 1.1$, $\beta = 1.1$.

0.8 gives rise to rather rough landscapes with fractal coastlines as shown in figure 13.9.

As discussed in the previous section, the superposition of a large number of nonfractal translation surfaces may well generate manifestly fractal surfaces, but the pictures presented so far involve only four layers (12 in figure 13.7), and with $\alpha > 1$ the height profiles $z(x, y)$ are rather smooth. These pictures are therefore nonfractal. What the effect of going to the limit of infinite spatial resolution (i.e., including infinitely high spatial frequencies) would be has not been studied in detail.

Weierstrass-Mandelbrot Coastlines

By changing the definition of the spectrum of allowed frequencies by replacing equation (13.3) by

$$f_j = b\, f_{j-1}\ , \quad \text{for}\quad j = 1, 2, \ldots\ ,$$

with $f_0 = 1$, we find that $\mathcal{Z}$ becomes the Weierstrass-Mandelbrot fractal function in equation (2.14) if we also choose the spectrum amplitude exponent to be given by $\beta = (2 - D)$.

Figure 13.10 shows a fractal landscape for which we have chosen the fundamental spatial frequency to be $f_0 = b = 4.7$. The amplitude spectrum

FIGURE 13.8: Landscape with perspective and curvature; $\alpha = 1.4$, $\beta = 1.1$.

exponent chosen is $\beta = 0.59$. The landscape is quite irregular and the fractal dimension for the coastline shown is found to be $D \simeq 1.35$. This should be compared to the expected fractal dimension for the Weierstrass-Mandelbrot curves given by $\beta = (2 - D)$, which here is $D = 1.405$ — an acceptable agreement considering the resolution used.

13.4 Random Addition Surfaces

The successive random addition algorithm introduced by Voss (1985b) was discussed in section 9.8, where it was used to generate fractional Brownian motion. The algorithm is, however, easily extended to higher dimensions as discussed by Voss. The landscapes shown in figure 13.11 were generated with this algorithm. We started by specifying the altitude $Z = 0$, on the four corners of the 1025×1025 lattice. The program uses a subroutine to generate an independent Gaussian variable ξ with zero mean and unit variance. In the first generation we simply generate one value for ξ and use it as the level at the central point (at $513,513$) of the lattice. In the next generation we first interpolate to find the elevation at the four points $(\frac{1}{4}, \frac{1}{4})$, $(\frac{3}{4}, \frac{1}{4})$, $(\frac{1}{4}, \frac{3}{4})$ and $(\frac{3}{4}, \frac{3}{4})$, using units for which the side of the lattice is set to 1. For example, the altitude at $(\frac{1}{4}, \frac{1}{4})$, is $Z(\frac{1}{4}, \frac{1}{4}) = \frac{1}{4}(Z(0,0) + Z(\frac{1}{2}, 0) + Z(\frac{1}{2}, \frac{1}{2}) + Z(0, \frac{1}{2}))$, i.e., the interpolated altitude is simply the

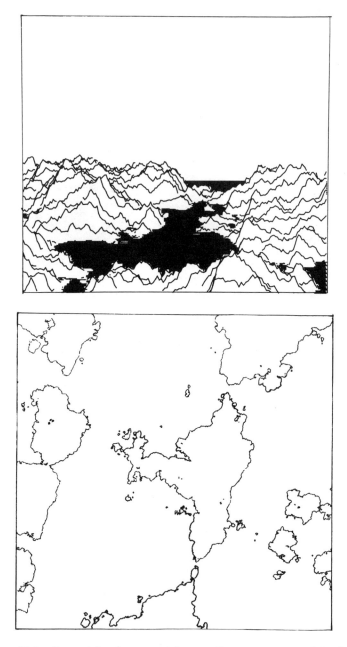

Figure 13.9: Fractal landscape with coastline map. $\alpha = 0.8$, $\beta = 2.0$. Coastline fractal dimension $D = 1.15$.

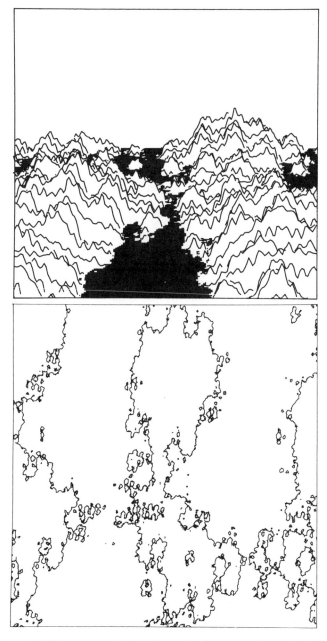

FIGURE 13.10: Weierstrass-Mandelbrot landscape. $\beta = 0.59$, $b = 4.7$. Coastline fractal dimension $D = 1.35$.

average of the altitude of the *diagonal* neighbors. The altitudes for the two neighbors on the rim [at $(0, \frac{1}{2})$ and $(\frac{1}{2}, 0)$] are taken to be the average of the altitudes of the corresponding corners of the rim. At this stage of the process we have specified the interpolated altitudes at thirteen positions, the five original locations, four new ones inside the region and four new positions on the rim. The next step is to use thirteen independent values of $\xi_{n=1}$, which are added to the elevations we have. The Gaussian random variable now has the variance (with $n = 1$) given by

$$\langle \xi_n^2 \rangle = \sigma_n^2 = r^{2nH} , \quad \text{with } r = 1/\sqrt{2} . \tag{13.5}$$

This is the same relation used in section 9.8, but with a scaling factor $r = 1/\sqrt{2}$, which is the change in distance between the old and new points here. This procedure is continued, adding in the next generation the points at $(\frac{1}{2}, \frac{1}{4})$, $(\frac{1}{4}, \frac{1}{2})$, $(\frac{3}{4}, \frac{1}{2})$ and $(\frac{1}{2}, \frac{3}{4})$. The altitudes at these locations are obtained as the average of the nearest-neighbor locations, i.e., the neighbors in directions parallel to the axes. Sites on the rim are again treated separately. In each generation this algorithm doubles the number of positions at which the altitudes are specified, and reduces the distance between the points by a factor $r = 1/\sqrt{2}$. The landscapes in figure 13.11 were obtained using 18 generations of this algorithm. The landscapes were generated using the *same* random numbers and therefore the three landscapes differ only in the value of the Hurst exponent chosen.

Sites on the rim are treated differently from the sites that are generated inside the region. The result of this boundary condition is that the landscapes are fractal, with a fractal dimension $D = 3 - H$, only on scales small compared with the dimension of the lattice.

The surfaces generated by this process are self-affine when all the points are treated on an equal footing. The x- and y-directions are different from the vertical z-direction, and therefore one has to distinguish between local and global fractal dimensions as discussed in chapter 10. A horizontal section of this self-affine surface defines coastlines like that shown in figure 13.12, which are self-similar fractals with a fractal dimension given by $D = 2 - H$, where H is the exponent used in the generation of the landscapes.

The fractal landscapes have a great deal of complexity and the graphics determine to a large extent how these landscapes are perceived and interpreted. For example, in figure 13.13, we also show the $H = 0.7$ landscape as seen from above and in a form that reminds one of clouds. Consider clouds that are large in horizontal extent and viewed from below. The formation of water droplets is a critical process and droplets are formed only when the water vapor supersaturation is above a certain level. If the

FIGURE 13.11: Fractal landscapes generated using Voss's successive random addition algorithm. The landscapes were generated on a 1025×1025 square grid. The variance of the additions was $\sigma^2 = (1/\sqrt{2})^{2nH}$ in the n-th generation for 18 generations. The Hurst exponents used were $H = 0.5$, 0.7 and 0.9, in the top, middle and bottom figures, respectively (Boger et al., 1987).

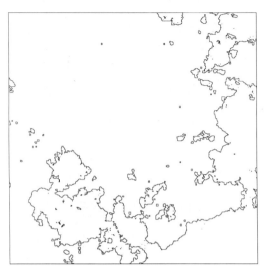

FIGURE 13.12: The coastline of the $H = 0.7$ landscape shown in figure 13.11. The coastline is self-similar with a fractal dimension $D = 2 - H = 1.3$ (Boger et al., 1987).

altitude of the landscape represents the determining factor for droplet formation (some combination of such factors as temperature and humidity), then we expect droplets only when the altitude is above a certain level. The clouds in figure 13.13 were made using a black background up to some level in the landscape, and from there on we used a gray scale from black to white proportional to the logarithm of the altitude. We find that the resulting pictures capture much of the texture and structure of clouds. For color versions of such cloud pictures see the insert following the Contents. Such clouds cannot, however, be incorporated into the pictures of landscapes since they are two-dimensional. Voss (1985b) has generated three-dimensional clouds that cast shadows and may be used to generate realistic scenes.

A close inspection of the landscapes shown in figure 13.11 shows that the algorithm used tends to produce rather sharp peaks. This is a feature of the algorithm that would be wiped out only after a very large number of generations. In the color plates following the Contents we also show landscapes generated using an algorithm in which the length scale is changed by a factor $r = 1/2$ in each generation. Starting with the altitudes at the four corners of the rim we interpolate these altitudes at the four positions $(\frac{1}{4}, \frac{1}{4}), (\frac{3}{4}, \frac{1}{4}), (\frac{3}{4}, \frac{3}{4})$ and $(\frac{1}{4}, \frac{3}{4})$. We then give random additions as before, interpolate again and so on. This algorithm has been used by Voss (1985b) and has also been discussed by Miller (1986). Landscapes generated using

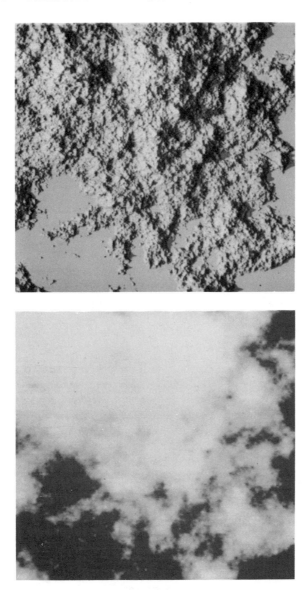

FIGURE 13.13: Fractal landscape with $H = 0.7$. The top figure is the landscape shown in figure 13.11 seen from above. The bottom figure renders this landscape as 'clouds.' The altitudes below sea level (see figure 13.12) are black. The gray scale used is proportional to the logarithm of the altitude above sea level. For a color version of the cloud see the insert following the Contents (Boger et al., 1987).

this algorithm have a different *lacunarity* and are quite satisfactory. As r is decreased the lacunarity of the landscape is increased and fewer fluctuation scales are visible.

13.5   Comments on Fractal Landscapes

Every landscape we have described may be considered to be generated by a simple process, which is controlled by the parameters α and β, β and b or r and H. To the extent that the reader finds the drawings acceptable as sketches of natural landscapes, this way of generating landscapes must be useful in the characterization of real landscapes. Clearly the graphics should be improved.

Landscapes of the type generated contain, of course, no information on the processes involved in the formation of the landscapes. However, they represent a very simple algorithmic description of landscapes. To describe an ellipse we need two numbers, and here to define a landscape we need three: α, β and the number of layers, or r, H and the number of generations. More complicated landscape geometries are easily generated. We might tilt the whole landscape to generate a coastal region, or we could add layers with different characteristics.

It is tempting to consider models in which a generated landscape is eroded or otherwise modified. However, at present we believe that the most important aspect of fractal landscapes is that they in fact are easy to specify and that they serve as simple geometrical tools for the description of complicated surfaces. Note that Voss's algorithm has the advantage that certain elevations (such as the rim) may be fixed in advance. Thus one may use his algorithm to perform *conditional* simulations in which information on the elevations at some locations is given. The 'landscapes' could also represent concentrations, temperatures or other quantities distributed over a surface. In fact, a nice example is provided by Hewett (1986), who used this method to generate distributions of porosity in a region, based on the information from a few well logs (see figure 8.11). Voss's algorithm has the further advantage that it can be iterated to any desired resolution. This method merges the simulation of landscapes with the analysis of observations — a most promising approach.

Chapter 14

Observations of Fractal Surfaces

In recent years many studies of the fractal structure of surfaces have appeared in the literature. Everything from protein surfaces to airport runways has been claimed to be fractal. The observations discussed use the full array of tools known to chemistry and physics. Generally speaking the observed fractal behavior does *not* cover many orders of magnitude in length scales, and the validity of the determined fractal dimensions may be questioned. Nevertheless, a very interesting set of observations have been analyzed and we discuss some of the recent results here.

14.1   Observed Surface Topography

Sayles and Thomas (1978a) have discussed and measured the surface roughness of objects ranging from supertanker hulls and concrete runways to hip joints and honed bearing raceways.

The height z of a surface is measured as a function of the distance x along some direction. With a large number of measurements all over the surface available, one may calculate the surface roughness as given by the variance

$$\sigma^2 = \langle z(x)^2 \rangle .$$

Here $\langle \ldots \rangle$ is the average over the set of observations or repeated observations of surface topography. The reference point in the vertical direction is chosen so that $\langle z(x) \rangle = 0$.

An important measure of the surface statistics is the correlation function defined by

$$C(\Delta x) = \langle z(x + \Delta x)z(x) \rangle.$$

For stationary surfaces the correlation function may be expressed in terms of the *power spectral density* $G(f)$ by a Fourier transform :

$$C(\Delta x) = \frac{1}{2} \int_{-\infty}^{\infty} G(f) \exp(2\pi i f \Delta x) df .$$

The spatial frequency f is related to the wavelength λ of the undulations on the surface by $f = 1/\lambda$. Physical systems have a finite extent L_{max}, and therefore a minimum spatial frequency $f_{min} = 1/L_{max}$. Therefore we may rewrite the correlation function in the form

$$C(\Delta x) = \int_{f_{min}}^{\infty} G(f) \exp(2\pi i f \Delta x) df .$$

Sayles and Thomas assume the following form for the power spectral density:

$$G(f) = k/f^2 , \tag{14.1}$$

and call the constant k the 'topethesy.' With this assumption the variance is given by

$$\sigma^2 = \langle z(x)^2 \rangle = \int_{f_{min}}^{\infty} G(f) df = k/f_{min} ,$$

i.e., we have $\sigma^2 = k L_0$, and the variance increases with the size of the surface as expected for Gaussian random processes.

In figure 14.1 we have reproduced their results. They have plotted $y = \log(\frac{1}{k}G(1/\lambda))$ versus $x = \log(\lambda)$. From equation (14.1) one expects a straight line with slope 2. They obtain a remarkable data collapse for 23 types of surfaces spanning 8 decades in surface wavelength. They suggest that the 'topethesy' k uniquely defines the statistical geometry of the random components of an isotropic surface for any given range of wavelength!

However, it should be noted that fitting the spectral density equation (14.1) to observations determines k, and by the scaling used this equation becomes $G(1/\lambda)/k = \lambda^2$. As pointed out by Berry and Hannay (1978), this amounts to taking raw data consisting of 23 short line segments with varying slopes scattered all over the log-log plot and translating them each along the vertical y-axis so that the segment lies as closely as possible to the line $y = 2x$. This procedure is guaranteed to yield a better-looking fit as the total range covered by the data increases.

Berry and Hannay (1978) comment that statistically isotropic surfaces which have no scale and whose height is well defined but nondifferentiable may indeed have spectra of the *fractal* form

$$G(f) = k/f^{\alpha} = k f^{-2H-1} . \tag{14.2}$$

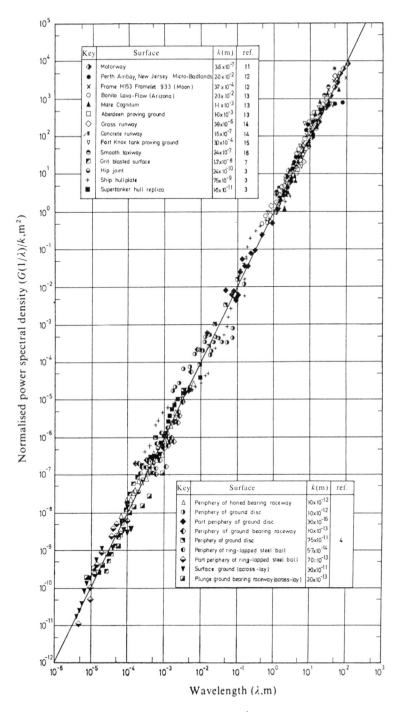

FIGURE 14.1: Scaled surface spectral power $\frac{1}{k}G(1/\lambda)$ versus surface height wavelength λ for different systems (Sayles and Thomas, 1978a).

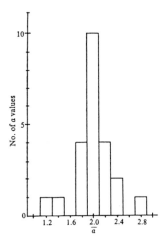

FIGURE 14.2: Histogram for the exponent α for the 23 sets of data in the previous figure (Sayles and Thomas, 1978b).

As discussed by Mandelbrot (1982, p. 353) the exponent H appearing here is the fractal codimension and is given in terms of the surface fractal dimension D by

$$D = 3 - H .$$

For *Brownian* surfaces — that is for ordinary Gaussian statistics — the equation (14.1) assumed by Sayles and Thomas is obtained since $H = 1/2$ and $D = 2.5$ for such surfaces. However, the parameter α must be considered a parameter of the fit, and it is found to be in the range 1.07 to 3.03, corresponding to a fractal dimension $D = (7 - \alpha)/2$ in the range 2 to 3. Sayles and Thomas (1978b) respond by refitting their data and presenting a plot of the obtained estimates of the spectrum parameter α as shown in figure 14.2. The values for α obtained cluster around the Gaussian value of 2, but cover the permissible range 1 to 3. We find this reasonable since we hardly expect the statistical properties of ball bearings and runways to be the same. Nevertheless, the analysis of Sayles and Thomas is interesting and its limitations should be tested critically on high-quality data.

Fractal Fracture Surfaces

When a piece of metal is fractured the fracture surface that is formed is rough and irregular. Mandelbrot et al. (1984) investigated the fractal structure of such surfaces. They studied fractured samples of 300-Grade Maraging steel samples by plating the fracture surfaces with nickel. The specimens were polished parallel to the plane of the fracture. 'Islands' of

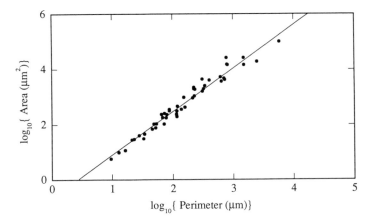

FIGURE 14.3: The perimeter-area relation for the fracture surface of 300-Grade Maraging steel. The line is a fit of $A \sim L^{2/D'}$ with $D' = 1.28$ (Mandelbrot et al., 1984).

steel surrounded by nickel appeared which, on subsequent polishing, grew and merged. The 'coastlines' or perimeter P and the area A of these islands were measured using a ruler or 'yardstick' $\delta = 1.5625\,\mu$m.

Fractal surfaces such as the fracture surfaces must scale differently in the plane of the fracture and in the direction perpendicular to the surface. Therefore a fracture surface is at best self-affine with a *local* fractal dimension D. However, the intersection of such a self-affine surface by a plane gives rise to coastlines which are indeed self-similar and have a fractal dimension $D' = D - 1$ (Mandelbrot, 1985; Voss, 1985a). Therefore we may use the *perimeter-area* relation (12.2) in the form

$$L(\delta) \sim \sqrt{A(\delta)}^{D'} \, . \tag{14.3}$$

In figure 14.3, we show the results obtained by Mandelbrot et al. (1984). A fit of equation (14.3) gives $D' = 1.28$, which implies that the fracture surface itself has a fractal dimension of $D = 2.28$, over a considerable range of length scales. Mandelbrot et al. also tested the fractal structure using a profile analysis. The fractured surface was sectioned to expose its profile and they determined the power spectral density $G(f)$ for the measured profiles. Using equation (14.2) H was determined and then they found the fractal dimension D of the surface by the relation

$$D = 3 - H = 1.26 \, ,$$

which is in good agreement with the previously determined value.

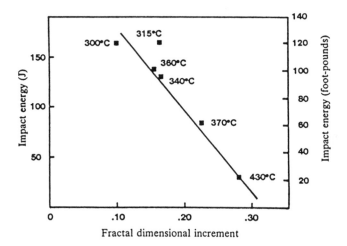

Figure 14.4: The relation between the observed surface fractal dimension D and the impact energy required to fracture a series of 300-Grade Maraging steel samples heat-treated at various temperatures (Mandelbrot et al., 1984).

Mandelbrot et al., in another interesting series of experiments, heat treated identical specimens of 300-Grade Maraging steel at different temperatures. They measured the impact energies required to fracture the samples and determined the surface fractal dimensions. In figure 14.4 we show their results. It is clear that the fractal dimensions, measured to be in the range $D = 2.1 - 2.28$, are approximately a linear function of the impact energy. The metallurgical basis for this relation is not understood, but after the discovery of the relation between the fracture energy and the fracture fractal dimension one at least has a handle on the surface topography.

14.2 D for Landscapes and Environmental Data

Burrough (1981) has analyzed a large number of environmental data and has obtained estimates of fractal dimensions for diverse properties as shown in figure 14.5. The various properties discussed are considered to be series of values z, sampled at regularly spaced positions x. The random function $z(x)$ corresponds to a set of points in the x, z-plane with a fractal dimension in the range $0 \leq D \leq 2$. The variance of increments defined by

$$V(\Delta x) = \langle [z(x + \Delta x) - z(x)]^2 \rangle$$

Property	Lag	D as lag $\to 0$
Soil—sodium content	15.2 m	1.7-1.9*
—stone content	15.2 m	1.1-1.8*
(both over four directions)		
Soil—thickness of cover loam	20 m	1.6*
Soil—electrical resistivity (4 directions)	1 m	1.4-1.6*
Surface of airport runway	30 cm	1.5†
Soil – mean cone index	~1 km	1.9‡
—silt + clay in 0-15 cm layer	~1 km	1.8‡
—mean diameter of surface stones	~1 km	1.8‡
—coarse sand fraction in 0-15 cm layer	~1 km	1.8‡
Vegetation cover	~1 km	1.6‡
Gold	Various	1.9*
Soil—phosphorus level	5 m	2.0‡
—pH	5 m	1.5‡
—potassium level	5 m	1.6‡
—bulk density	5 m	1.5‡
—0.1 bar water	5 m	1.5‡
Iron ore in rocks		
—chlorite	15 μm	1.6*
—quartz	15 μm	1.9*
—quartz	5 cm	1.6*
—iron	5 cm	1.5*
—iron (E–W)	100 m	1.7*
—iron (N–S)	100 m	1.8*
—iron (E–W)	500 m	1.6*
—iron (N–S)	500 m	1.9*
Sea anemones	10 cm	1.6§
Rainfall	1 km	1.7*
Iron ore	3 m	1.4*
Groundwater levels		
Piezometer 1	1 day	1.6*
2	1 day	1.7*
3	1 day	1.8*
4	1 day	1.3*
Oil grades	60 cm	1.7*
Copper grades	2 m	1.7*
Topographic heights	10 m	1.5*
Soil—sand content	10 m	1.6-1.8*
—pH	10 m	2.0*
Crop yields	1-1,000 m	1.6-1.8‡
Water table depth	250 m	1.6*

FIGURE 14.5: Estimated D for various environmental series. For an explanation of the symbols see the text (Burrough, 1981).

has a position dependence given by $V(\Delta x) \sim |\Delta x|^{2H}$, as discussed by Mandelbrot (1982, p. 353). Therefore the fractal dimension $D = 2 - H$ may be estimated from a log-log plot of the variance of increments. This method has been used for the entries marked * in figure 14.5. The power spectrum is given by equation (14.2), and the fractal dimension is again obtained from an estimate of the codimension H, by $D = 2 - H$; the resulting estimates are marked †. Related estimates involving block variances are marked with ‡, and estimates based on covariances are marked §.

The values for D reported fall in the range 1.4 to 2.0. The high values of D reported in the table for some soil and geological data seem to question the wisdom of interpolating mapping in certain instances, and it would seem worthwhile to use D as a guide to how further mapping and interpolation should proceed. We find this analysis interesting but in order to assess its validity one has to examine critically the original data, and in particular one must find the range over which there is no length scale. In fact experience from research on phase transitions for which scale invariance is well documented and understood indicates that reliable estimates of exponents can only be obtained when the data extend over at least three decades — this stringent criterion has not been met so far in the discussion of fractal surfaces or in the environmental data.

14.3   Molecular Fractal Surfaces

Surfaces may be rough and even fractal down to the molecular size range. The area of surfaces may be measured — in the spirit of the Hausdorff-Besicovitch definition — by adsorbing molecules of different size on the surface and 'counting' their number. Surface areas are usually determined by measuring *adsorption isotherms.* One measures the number of moles n of the molecules that are *adsorbed* on the surface as a function of the pressure P at a given temperature T:

$$n = f_T(P) \ .$$

One method for the determination of n makes use of pressure-volume measurements to determine the amount of gas before and after exposure to the adsorbent. A second general type of procedure is to determine n by a direct weighing of the amount of adsorption. For the adsorption of molecules from solution many special techniques are used. For a good general discussion of surface adsorption see Adamson (1982).

The *Langmuir isotherm*, often used in the interpretation of adsorption isotherms, has the form

$$n = \frac{n_m bP}{1 + bP} \ . \tag{14.4}$$

Here n_m denotes the number of moles adsorbed at the monolayer point. From such isotherms the n_m is determined. The surface area Σ is given by

$$\Sigma = n_m N_A \sigma_0 \ ,$$

where σ_0 is the area occupied by an adsorbed molecule, and N_A is Avogadro's number. For many substances the adsorption area is well known

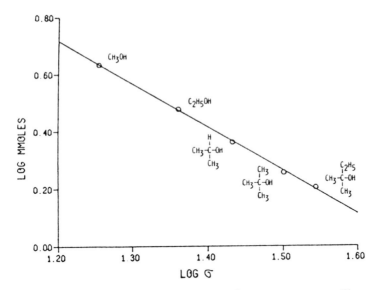

FIGURE 14.6: Measured monolayer mole numbers n on porous silica gel as a function of molecule cross-section $\sigma \, (\text{Å}^2)$. $D = 3.02 \pm 0.06$ (Pfeifer et al., 1984).

and to a large extent independent of the substrate. In many practical applications nitrogen is used, and the accepted value of the area per site is $\sigma_0 = 16.2 \, \text{Å}^2$. Another widely used method used to interpret adsorption isotherms has been introduced by Brunauer et al. (1938): the so-called *BET isotherm*, which is useful for more complicated adsorption isotherms.

Observation of Molecular Fractal Surfaces

In a series of papers Avnir, Pfeifer and Farin (Avnir et al., 1983, 1984; Avnir and Pfeifer, 1983; Pfeifer and Avnir, 1983; Pfeifer et al., 1983, 1984) have discussed the surface areas determined by adsorption isotherms and concluded that many substances are *fractal* and characterized by a surface fractal dimension in the range $2 \le D \le 3$. The specific surface of the sample depends on the *size* of the molecules used. They note that with a length scale δ given in terms of the adsorption area

$$\sigma_0 = \delta^2 \,,$$

the amount adsorbed on a sample with a fractal surface must have the form

$$n \sim \delta^{-D} = \sigma^{-D/2} \,, \tag{14.5}$$

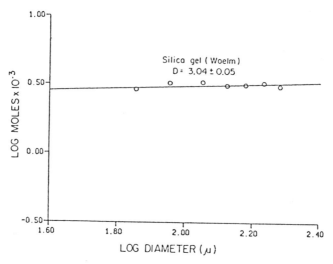

FIGURE 14.7: Measured monolayer mole numbers n of tertiary amyl alcohol as a function of adsorbent particle diameter $(2R)$ (μm) (Pfeifer et al., 1984).

consistent with equations (2.3) and (2.4). As an example of this type of behavior consider figure 14.6, which shows the results obtained by Pfeifer et al. (1984) for the adsorption of spherical alcohols from toluene on porous silica gel. The mole numbers observed follow the relation (14.5) with $D = 3.02 \pm 0.06$ over the yardstick range $\sigma_0 = 18$ to $35\,\text{Å}^2$ or δ in the range 4.2 to 5.9 Å. This is an extreme value of the fractal dimension of a surface. The surface is so crumpled and porous that it practically fills the volume. It follows that in thermodynamic terms the surface terms contribute as much as the volume terms. A monolayer of adsorbed molecules in this system amounts to a bulk phase interrupted by an increasing number of ever smaller voids. This remarkably high value for the surface fractal dimension should be investigated over a larger range of yardsticks.

Recently Rojanski et al. (1986) investigated mesoporous silica gel by adsorption, electronic energy transfer and small-angle X-ray scattering. All these different methods gave the result that the surface is extremely rough and irregular, with a fractal dimension $D \simeq 3$.

Pfeifer and Avnir (1983) have introduced a method that permits a large extension of the effective yardstick range. Consider geometrically similar particles. Let R be the radius of the smallest sphere needed to enclose such a particle. Then by the same arguments used for the derivation of the perimeter-area relation (12.2) we obtain the area-volume relation

$$A \sim V^{D/3} . \tag{14.6}$$

Here the particle volume V is given by $V \sim R^3$. For a given sample volume $V_S \gg V$, filled with the adsorbent, the number of particles N_S in the sample will increase with decreasing radius of the particles as $N_S \sim R^{-3}$. Combining this with equation (14.6) it follows that the surface area Σ — with a fixed yardstick δ — will change with the radius of the particles as

$$\Sigma \sim n \sim R^{D-3} . \tag{14.7}$$

The monolayer mole number n measured for molecules of a given size is proportional to Σ as indicated. With this method of analysis Pfeifer and Avnir have extended the range of the results in figure 14.6 as shown in figure 14.7. Thus effectively they have obtained $D = 3.04 \pm 0.05$ over the yardstick range 35 to 256 $\overset{\circ}{A}^2$.

A collection of fractal dimensions determined by Avnir et al. (1984) is shown in figure 14.8, which is reproduced from their paper. This is a remarkable list of surface fractal dimensions ranging from $D = 2$ all the way to $D = 3$. Clearly these results warrant further research. In particular, the range over which the surface structure is measured should be extended into the regions where both the lower- and the high-end cutoff scales are reached. Again, experience from the work on phase transitions has shown that exponents determined from data that do not extend over at least three decades are unreliable.

We expect that the realization that molecular fractal surfaces exist will have a significant impact on many fields that relate to surfaces, such as catalysis, wetting and powder technology. This will certainly become a very active field of research.

Porosity with Fractal Properties

In a very interesting paper Bale and Schmidt (1984) proposed that microscopic porosity may be fractal. They gave a derivation of an expression for the X-ray scattering intensity from fractal pore surfaces,

$$S(q) \sim q^{(D-6)} , \tag{14.8}$$

where the scattering vector q is given by the same expression as in equation (3.4). From observations of small-angle scattering in Beulah lignite coal, they concluded that the pore space surface is fractal and has the fractal dimension $D = 2.56 \pm 0.03$ over almost two orders of magnitude in the scattering vector (figure 14.9). Clearly more small-angle scattering work is needed in order to obtain better insight into the fractal properties of porous systems.

Fractal dimension*	Found in	At the range (Å^2)†
High		
2.91 ± 0.02	Upper Columbus dolomitic rock from Bellevue, Ohio	20–47,000
2.97 ± 0.01	Goodland high calcium rock, from Idabel, Okla.	20–47,000
2.88 ± 0.02	Granitic rock from SHOAL nuclear test site, Nevada	16–16,500
2.73 ± 0.05	Igneous rock sample from SHOAL site	14–14,300
2.71 ± 0.14	Granular activated carbon–Tsurumi HC-8, from coconut shell	(16–37)
2.80 ± 0.16	Granular activated carbon—Fujisawa B-CG, from coconut shell	(16–37)
2.90 ± 0.01	Carbonate rock from groundwater test well, Yucca Flat, Nevada	16–16,500
2.92 ± 0.02	Soil (kaolinite, trace halloysite)	150–16,500
2.94 ± 0.04	Porous silicic acid	(16–34)
2.79 ± 0.03	Activated alumina grade F-20 (Alcoa Corp.)	16–45,100
2.78 ± 0.21	Charcoal (BDH) of animal origin	1,400–180,000
2.67 ± 0.16	Porous coconut charcoal (Standard Chemical Co., Montreal)	(16–47)
Medium		
2.57 ± 0.04	Porous α-FeOOH pigment for magnetic tapes	16–980
2.35 ± 0.11	Crushed Corning 0010 lead glass	21–14,900
2.52 ± 0.07	Coal mine dust from Western Pennsylvania	16–180
2.33 ± 0.08	Coal mine dust from Western Pennsylvania	16–270
2.25 ± 0.09	Carbon black	(16–71)
2.54 ± 0.12	Slightly porous coconut charcoal (Standard Chemical Co., Montreal)	(16–47)
2.30 ± 0.07	Slightly porous coconut charcoal (Standard Chemical Co., Montreal)	(16–47)
2.63 ± 0.03	Mosheim high calcium from Stephens City, Va.	20–47,000
2.58 ± 0.01	Niagara (Guelph) dolomite from Woodville, Ohio	20–47,000
2.46 ± 0.11	Glassy melted rock from Rainier nuclear zone, Nevada	14–14,300
2.29 ± 0.06	Soil (mainly feldspar quartz and limonite)	150–21,800
Low		
2.02 ± 0.06	Aerosil—nonporous fumed silica (Degussa)	16–529
2.15 ± 0.06	Snowit—ground fine Belgian quartz glass of high purity	16–10,600
2.14 ± 0.06	Madagascar quartz from Thermal Syndicate	16–1,850
1.95 ± 0.04	Periclase—electrically fused and crushed magnesite	16–720
2.02 ± 0.05	Synthetic faujasite ($Na_2O \cdot Al_2O_3 \cdot 2,67SiO_2 \cdot mH_2O$) (Linde Air Products)	(16–68)
2.07 ± 0.01	Graphite—Vulcan 3G (2700) (National Physical Lab., Teddington, UK)	(16–178)
2.04 ± 0.16	Graphon—partially graphitized carbon black formed by heating MPC black to 3,200 °C	(15–41)
2.13 ± 0.16	Graphon—graphitized carbon black (Cabot Corp.)	1,400–180,000
2.04 ± 0.04	Active, (nonporous), coconut charcoal (Standard Chemical Co., Montreal)	(16–47)
1.97 ± 0.02	Active, (nonporous), coconut charcoal (Standard Chemical Co., Montreal)	(16–47)
2.16 ± 0.04	Iceland spar, massive calcite from Chihuahua, Mexico	20–47,000

FIGURE 14.8: Fractal dimensions of surfaces measured by molecular adsorption (Avnir et al., 1984).

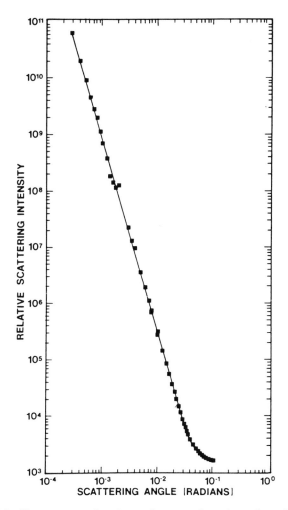

FIGURE 14.9: X-ray scattering intensity as a function of angle for Beulah lignite (Bale and Schmidt, 1984).

Wong (1985) discussed the theory of small-angle scattering and pointed out that these data should be interpreted as evidence of rough pore surfaces which are not fractals. Wong et al. (1986), using small-angle neutron scattering, found that sandstones and shale have fractal internal surfaces with D in the range 2.55 to 2.96 depending on the type of rock investigated. They identified clay as the origin of such structures.

Katz and Thompson (1985) investigated porous sandstones using a scanning electron microscope. They measured fractal parameters on rock

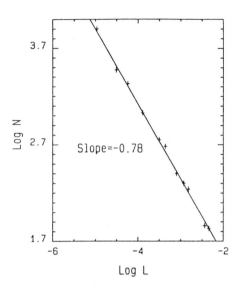

FIGURE 14.10: The number $N(L)$ of geometric features of size L per unit length versus the size L (cm) for Coconino sandstone (Katz and Thompson, 1985).

fracture surfaces utilizing the secondary-electron emission from the scanning electron beam. At each magnification the structure detected by the electron beam along a linear trace is limited by the microscope resolution; for an isotropic fractal this limits the depth of field. The scanning electron microscope depth of field decreases at increasing magnification such that it is always smaller than the depth of the rock interface. Therefore in effect the intensity of the secondary-emission electrons has features that come from the intersection of a line with the pore space–matrix interface. For a volume fractal with a surface fractal dimension D, the intersection with a line is a set of points with dimension $D' = D - 2$. Katz and Thompson hypothesized that, at a given magnification, the number of 'features' $N(L)$ per unit length resolved at a given magnification as a function of the length scale L should satisfy the relation $N(L) \sim L^{-(D-2)}$. As shown in figure 14.10 their observations are nicely fitted by this form and the fractal dimension estimated is $D = 2.78$. In table 14.1 we give their results for various specimens.

The fractal structure of the pore space has a lower cutoff $l_1 \sim 20\,\text{Å}$ and an upper cutoff l_2 above which the power-law dependence on L no longer holds. Katz and Thompson proposed the following relation for the porosity ϕ:

$$\phi = (l_1/l_2)^{3-D} .$$

(14.9)

Sample	Fractal dimension	l_2 (μm)	Porosity (%) Calculated	Measured
Tight gas sand #965	2.57	2.5	4.7	5.3–5.6
Tight gas sand #466	2.68	6	7.6	6.9–7.6
Coconino	2.78	98	10	11–12.5
Navajo	2.81	50	15	16.4
St. Peter's	2.87	50	27	24–28

TABLE 14.1: The fractal dimension of the pore surface measured using secondary-electron emission from sand and sandstones. The porosities have been measured, and the calculated values have been estimated using $\phi = (l_1/l_2)^{3-D}$ (Katz and Thompson, 1985).

They have used this relation to estimate ϕ from the observed upper length scale l_2, and find a satisfactory agreement. However, they give no argument for their postulated relation, which cannot hold in general. Recently this point has been criticized by Roberts (1986), and Katz and Thompson (1986) have answered. Again more research is needed.

References

ADAMSON A. W. 1982. *Physical Chemistry of Surfaces.* (4th edition, J. Wiley, New York).

AHARONY A. 1985. Anomalous diffusion on percolating clusters. in *Scaling Phenomena in Disordered Systems* (editors R. Pynn & A. Skjeltorp, Plenum Press, New York, pp. 289–300).

AHARONY A. 1986. Percolation. in *Directions in Condensed Matter Physics* (editors G. Grinstein & G. Mazenko, World Scientific, Singapore, pp. 1–50).

AHARONY A. 1987. Multifractality on percolation clusters. in *Time-Dependent Effects in Disordered Materials* (editors R. Pynn & T. Riste, Plenum Press, New York, pp. 163–171).

AHARONY A., & STAUFFER D. 1987. Percolation. *Encyclopedia of Physical Science and Technology* **10**, 226–244.

AHARONY A., GEFEN Y., KAPITULNIK A., & MURAT M. 1985. Fractal eigen-dimensionalities for percolation clusters. *Phys. Rev. B* **31**, 4721–4722.

AMITRANO C., CONIGLIO A., & DI LIBERTO F. 1986. Growth probability distribution in kinetic aggregation processes. *Phys. Rev. Lett.* **57**, 1016–1019.

AVNIR D., & PFEIFER P. 1983. Fractal dimension in chemistry. An intensive characteristic of surface irregularity. *Nouv. J. Chim.* **7**, 71–72.

AVNIR D., FARIN D., & PFEIFER P. 1983. Chemistry in noninteger dimensions between two and three. II. Fractal surfaces of adsorbents. *J. Chem. Phys.* **79**, 3566–3571.

AVNIR D., FARIN D., & PFEIFER P. 1984. Molecular fractal surfaces. *Nature* **308**, 261–263.

BADII R., & POLITI A. 1984. Hausdorff dimension and uniformity of strange attractors. *Phys. Rev. Lett.* **52**, 1661–1664.

BADII R., & POLITI A. 1985. Statistical description of chaotic attractors: The dimension function. *J. Stat. Phys.* **40**, 725–750.

BAK P. 1986. The devil's staircase. *Phys. Today* **39**, 38–45.

BALE H. D., & SCHMIDT P. W. 1984. Small-angle X-ray-scattering investigation of submicroscopic porosity with fractal properties. *Phys. Rev. Lett.* **53**, 596–599.

BARNSLEY M. F., & SLOAN A. D. 1988. A better way to compress images. *Byte* **13**, 215–223.

BEN-JACOB E., GODBEY R., GOLDENFELD N. D., KOPLIK J., LEVINE H., MUELLER T., & SANDER L.M. 1985. Experimental demonstration of the role of anisotropy in interfacial pattern formation. *Phys. Rev. Lett.* **55**, 1315–1318.

BENSIMON D., KADANOFF L. P., LIANG S., SHRAIMAN B., & TANG C. 1986a. Viscous flows in two dimensions. *Rev. Mod. Phys.* **58**, 977–999.

BENSIMON D., JENSEN M. H., & KADANOFF L. P. 1986b. Renormalization-group analysis of the global structure of the period doubling attractor. *Phys. Rev. A* **33**, 3622–3624.

BENZI R., PALADIN G., PARISI G., & VULPANI A. 1984. On the multifractal nature of fully developed turbulence and chaotic systems. *J. Phys. A* **17**, 3521–3531.

BERG H. C. 1983. *Random Walks in Biology.* (Princeton University Press, Princeton, New Jersey).

BERRY M. V., & HANNAY J. H. 1978. Topography of random surfaces. *Nature* **273**, 573.

BERRY M. V., & LEWIS Z. V. 1980. On the Weierstrass-Mandelbrot fractal function. *Proc. R. Soc. London A* **370**, 459–484.

BILLINGSLEY P. 1965. *Ergodic Theory and Information.* (J. Wiley, New York).

BILLINGSLEY P. 1983. The singular function of bold play. *Am. Sci.* **71**, 392–397.

BLUMENFELD R., MEIR Y., B. HARRIS A., & AHARONY A. 1986. Infinite set of exponents describing physics on fractal networks. *J. Phys. A* **19**, L791–L796.

BLUMENFELD R., MEIR Y., AHARONY A., & HARRIS A. B. 1987. Resistance fluctuations in randomly diluted networks. *Phys. Rev. B***35**, 3524–3535.

BOGER F., FEDER J., & JØSSANG T. 1987. Fractal landscapes generated using Voss's successive random addition algorithm. *Report Series, Cooperative Phenomena Project, Department of Physics, University of Oslo*, **87-15**, 1–11.

BRADY R. M., & BALL R. C. 1984. Fractal growth of copper electrodeposits. *Nature* **309**, 225–229.

BROADBENT S. R., & HAMMERSLEY J. M. 1957. Percolation processes I. Crystals and mazes. *Proc. Cambridge Philos. Soc.* **53**, 629–641.

BROWN R. 1828. On the existence of active molecules in organic and inorganic bodies. *Phil. Mag.* **4**, 162–173.

BRUNAUER S., EMMETT P. H., & TELLER E. 1938. Adsorption of gases in multimolecular layers. *J. Am. Chem. Soc.* **60**, 309–319.

BUKA A., KERTÉSZ J., & VICSEK T. 1986. Transitions of viscous fingering patterns in nematic liquid crystals. *Nature* **323**, 424–425.

BURROUGH P. A. 1981. Fractal dimensions of landscapes and other environmental data. *Nature* **294**, 240–242.

CHANDLER R., KOPLIK J., LERMAN K., & WILLEMSEN J. F. 1982. Capillary displacement and percolation in porous media. *J. Fluid Mech.* **119**, 249–267.

CHEN J. D. 1987. Pore-scale difference between miscible and immiscible viscous fingering in porous media. *AICE J.* **33**, 307–311.

CHEN J. D., & WILKINSON D. 1985. Pore-scale viscous fingering in porous media. *Phys. Rev. Lett.* **55**, 1892–1895.

CHUOKE R. L., VAN MEURS P., & VAN DER POEL C. 1959. The instability of slow, immiscible, viscous liquid-liquid displacements in permeable media. *Trans. Metall. Soc. of AIME* **216**, 188–194.

CLÉMENT E., BAUDET C., & HULIN J. P. 1985. Multiple scale structure of nonwetting fluid invasion fronts in 3D model porous media. *J. Phys. Lett.* **46**, L1163–L1171.

CONIGLIO A. 1981. Thermal phase transition of the dilute s-state Potts and n-vector models at the percolation threshold. *Phys. Rev. Lett.* **46**, 250–253.

CONIGLIO A. 1982. Cluster structure near the percolation threshold. *J. Phys. A* **15**, 3829–3844.

COURTENS E., & VACHER R. 1987. Structure and dynamics of fractal aerogels. *Z. Phys. B* **68**, 355–361.

DACCORD G. 1987. Chemical dissolution of a porous medium by a reactive fluid. *Phys. Rev. Lett.* **58**, 479–482.

DACCORD G., & LENORMAND R. 1987. Fractal patterns from chemical dissolution. *Nature* **325**, 41–43.

DACCORD G., NITTMANN J., & STANLEY H. E. 1986. Fractal viscous fingers: Experimental results. in *On Growth and Form* (editors H. E. Stanley & N. Ostrowsky, Martinus Nijhoff, Dordrecht, pp. 203–210).

DE ARCANGELIS L., REDNER S., & CONIGLIO A. 1985. Anomalous voltage distribution of random resistor networks and a new model for the backbone at the percolation threshold. *Phys. Rev. B* **31**, 4725–4727.

DE GENNES P. G. 1976. La percolation: Un concept unificateur. *La Recherche* **7**, 919–927.

DE GENNES P. G., & GUYON E. 1978. Lois genérales pour l'injection d'un fluide dans un milieu poreux aléatoire. *J. Mec.* **17**, 403–432.

DEGREGORIA A. J., & SCHWARTZ L. W. 1987. Saffman-Taylor finger width at low interfacial tension. *Phys. Rev. Lett.* **58**, 1742–1744.

DEN NIJS M. P. M. 1979. A relation between the temperature exponents of the eight-vertex and q-state Potts model. *J. Phys. A* **12**, 1857–1868.

DEUTSCHER G., ZALLEN R., & ADLER J., editors.1983. *Percolation Structures and Processes. Ann. Isr. Phys. Soc.* **5**.

DIAS M. M., & WILKINSON D. J. 1986. Percolation with trapping. *J. Phys. A* **19**, 3131–3146.

DUBINS L. E., & SAVAGE L. J. 1960. Optimal gambling systems. *PNAS* **46**, 1597–1598.

EINSTEIN A. 1905. Über die von der molekularkinetischen Theorie der Wärme geforderte Bewegung von in ruhenden Flüssigkeiten suspendierten Teilchen. *Ann. Phys.* **322**, 549–560.

ENGELBERTS W. F., & KLINKENBERG L. J. 1951. Laboratory experiments on the displacement of oil by water from packs of granular material. *Petr. Congr. Proc. Third World*, 544–554.

ENGLMAN R., & JAEGER Z., editors 1986. *Fragmentation Form and Flow in Fractured Media. Ann. Isr. Phys. Soc.* **8**.

ESSAM J. W. 1980. Percolation theory. *Rep. Prog. Phys.* **43**, 833–912.

EVANS D. C., & ATHAY R. J., editors 1986. *Computer Graphics.* (Siggraph 1986 Conference Proceedings, Computer Graphics **20** number 4, August 1986).

FALCONER K. J. 1985. *The Geometry of Fractal Sets.* (Cambridge University Press, Cambridge).

FAMILY F. & LANDAU D. P., editors. 1984. *Aggregation and Gelation.* (North-Holland, Amsterdam).

FEDER J., & JØSSANG T. 1985. A reversible reaction limiting step in irreversible immunoglobulin aggregation. in *Scaling Phenomena in Disordered Systems* (editors R. Pynn & A. Skjeltorp, Plenum Press, New York, pp. 99–131).

FEDER J., JØSSANG T., & ROSENQVIST E. 1984. Scaling behavior and cluster fractal dimension determined by light scattering from aggregating proteins. *Phys. Rev. Lett.* **53**, 1403–1406.

FEDER J., JØSSANG T., MÅLØY K. J., & OXAAL U. 1986. Models of viscous fingering. in *Fragmentation Form and Flow in Fractured Media* (editors R. Englman & Z. Jaeger, *Ann. Isr. Phys. Soc.* **8**, pp. 531–548).

FEIGENBAUM M. J., JENSEN M. H., & PROCACCIA I. 1986. Time ordering and the thermodynamics of strange sets: Theory and experimental tests. *Phys. Rev. Lett.* **57**, 1503–1506.

FELLER W. 1951. The asymptotic distribution of the range of sums of independent variables. *Ann. Math. Stat.* **22**, 427–432.

FELLER W. 1968. *An Introduction to Probability Theory and Its Applications.* (vol 1., 3rd edition, John Wiley, London).

FJØRTOFT R. 1982. A study of the wave climate in the Norwegian Sea. Algorithms in Markov models for deriving probabilities of certain events. *Geophys. Norv.* **32**, 45–76.

FRISCH U., & PARISI G. 1985. On the singularity structure of fully developed turbulence. in *Turbulence and Predictability in Geophysical Fluid Dynamics and Climate Dynamics* (editors M. Ghil, R. Benzi & G. Parisi, North-Holland, New York, pp. 84–88).

FRØYLAND J., FEDER J., & JØSSANG T. 1988. The fractal statistics of ocean waves. *To be published.*

GARDNER M. 1976. Mathematical games. *Sci. Am.* **235**, 124–133.

GEFEN Y., MANDELBROT B. B., & AHARONY A. 1980. Critical phenomena on fractal lattices. *Phys. Rev. Lett.* **45**, 855–858.

GEFEN Y., AHARONY A., ALEXANDER S. 1983. Anomalous diffusion on percolating clusters. *Phys. Rev. Lett.* **50**, 77–80

GLAZIER J. A., JENSEN H. H., LIBCHABER A., & STAVANS J. 1986. Structure of Arnold tongues and the $f(\alpha)$ spectrum for period doubling: Experimental results. *Phys. Rev. A* **34**, 1621–1624.

GORDON J. M., GOLDMAN A. M., MAPS J., COSTELLO D., TIBERIO R., & WHITEHEAD B. 1986. Superconducting-normal phase boundary of a fractal network in a magenetic field. *Phys. Rev. Lett.* **56**, 2280–2283.

GRASSBERGER P. 1983. Generalized dimensions of strange attractors. *Phys. Lett. A* **97**, 227–230.

GRASSBERGER P., & PROCACCIA I. 1983. Measuring the strangeness of strange attractors. *Physica* **9**, 189–208.

GROSSMAN T., & AHARONY A. 1986. Structure and perimeters of percolation clusters. *J. Phys. A* **19**, L745–L751.

GROSSMAN T., & AHARONY A. 1987. Accessible external perimeters of percolation clusters. *J. Phys. A* **20**, L1193–L1201.

HABERMANN B. 1960. The efficiency of miscible displacement as a function of mobility ratio. *Soc. Pet. Eng. AIEME J.* **219**, 264–272.

HACK J. T. 1957. Studies of longitudinal stream profiles in Virginia and Maryland. *U. S. Geol. Surv. Prof. Pap.* **294-B**, 45–97.

HALSEY T., MEAKIN P., & PROCACCIA I. 1986a. Scaling structure of the surface layer of diffusion-limited aggregates. *Phys. Rev. Lett.* **56**, 854–857.

HALSEY T. C., JENSEN M. H., KADANOFF L. P., PROCACCIA I., & SHRAIMAN B. I. 1986b. Fractal measures and their singularities: The characterization of strange sets. *Phys. Rev. A* **33**, 1141–1151.

HAMMERSLEY J. M. 1983. Origins of percolation theory. in *Percolation Structures and Processes* (editors G. Deutscher, R. Zallen & J. Adler, *Ann. Isr. Phys. Soc.* **5**, pp. 47–57).

HAVLIN S., & NOSSAL R. 1984. Topological properties of percolation clusters. *J. Phys. A* **17**, L427–L432.

HAYAKAWA Y., SATO S., & MATSUSHITA M. 1987. Scaling structure of the growth-probability distribution in diffusion-limited aggregation processes. *Phys. Rev. A* **36**, 1963–1966.

HELE-SHAW H. S. 1898. The flow of water. *Nature* **58**, 34–36.

HENTSCHEL H. G. E., & PROCACCIA I. 1983a. Fractal nature of turbulence as manifested in turbulent diffusion. *Phys. Rev. A* **27**, 1266–1269.

HENTSCHEL H. G. E., & PROCACCIA I. 1983b. The infinite number of generalized dimensions of fractals and strange attractors. *Physica* **8**, 435–444.

HENTSCHEL H. G. E., & PROCACCIA I. 1984. Relative diffusion in turbulent media: The fractal dimension of clouds. *Phys. Rev. A* **29**, 1461–1470.

HERRMANN H. J. 1986. Geometrical cluster growth models and kinetic gelation. *Phys. Rep.* **136**, 154–227.

HERRMANN H. J., & STANLEY H. E. 1984. Building blocks of percolation clusters: Volatile fractals. *Phys. Rev. Lett.* **53**, 1121–1124.

HEWETT T. A. 1986. Fractal distributions of reservoir heterogeneity and their influence on fluid transport. Paper SPE 15386 presented at the 61st Annual Technical Conference and Exhibition of the Society of Petroleum Engineers, New Orleans, Oct. 5–8, 1986.

HINRICHSEN E., FEDER J., & JØSSANG T. 1987. DLA growth from a line. *Report Series, Cooperative Phenomena Project, Department of Physics, University of Oslo* **87-11**, 1–21.

HOMSY G. M. 1987. Viscous fingering in porous media. *Ann. Rev. Fluid Mech.* **19**, 271–311.

HORTON R. E. 1945. Erosional development of streams and their drainage basins; hydrophysical approach to quantitative morphology. *Geol. Soc. Am. Bull.* **56**, 275–370.

HORVÁTH V., VICSEK T., & KERTÉSZ J. 1987. Viscous fingering with imposed uniaxial anisotropy. *Phys. Rev. A* **35**, 2353–2356.

HURST H. E. 1951. Long-term storage capacity of reservoirs. *Trans. Am. Soc. Civ. Eng.* **116**, 770–808.

HURST H. E., BLACK R. P., & SIMAIKA Y. M. 1965. *Long-Term Storage: An Experimental Study.* (Constable, London).

JENSEN M. H., BAK P., & BOHR T. 1983. Complete devil's staircase, fractal dimension, and universality of mode-locking structure in circle maps. *Phys. Rev. Lett.* **50**, 1637–1639.

JENSEN M. H., BAK P., & BOHR T. 1984a. Transition to chaos by interaction of resonances in dissipative systems I. Circle maps. *Phys. Rev. A* **30**, 1960–1969.

JENSEN M. H., BAK P., & BOHR T. 1984b. Transition to chaos by interaction of resonances in dissipative systems II. Josephson junctions, charge-density waves, and standard maps. *Phys. Rev. A* **30**, 1970–1981.

JENSEN M. H., KADANOFF L. P., LIBCHABER A., PROCACCIA I., & STAVANS J. 1985. Global universality at the onset of chaos: Results of a forced Rayleigh-Bénard experiment. *Phys. Rev. Lett.* **55**, 2798–2801.

JENSEN M. H., LIBCHABER A., PELCÉ P., & ZOCCHI G. 1987. Effect of gravity on the Saffman-Taylor meniscus: Theory and experiment. *Phys. Rev. A* **35**, 2221–2227.

JØSSANG T., FEDER J., & ROSENQVIST E. 1984. Heat aggregation kinetics of human IgG. *J. Chem. Phys.* **120**, 1–30.

JULLIEN R., & BOTET R. 1987. *Aggregation and Fractal Aggregates.* (World Scientific, Singapore).

KAPITULNIK A., AHARONY A., DEUTSCHER G., & STAUFFER D. 1983. Self similarity and correlations in percolation. *J. Phys. A* **16**, L269–L274.

KATZ A. J., & THOMPSON A. H. 1985. Fractal sandstone pores: Implications for conductivity and pore formation. *Phys. Rev. Lett.* **54**, 1325–1328.

KATZ A. J., & THOMPSON A. H. 1986. Katz and Thompson respond. *Phys. Rev. Lett.* **56**, 2112.

KATZEN D., & PROCACCIA I. 1987. Phase transitions in the thermodynamic formalism of multifractals. *Phys. Rev. Lett.* **58**, 1169–1172.

KIRKPATRICK S. 1973. Percolation and conduction. *Rev. Mod. Phys* **45**, 574–588.

KJEMS J., & FRELTOFT T. 1985. Neutron and X-ray scattering from aggregates. in *Scaling Phenomena in Disordered Systems* (editors R. Pynn & A. Skjeltorp, Plenum Press, New York, pp. 133–140).

LAIDLAW D., MACKAY G., & JAN N. 1987. Some fractal properties of the percolating backbone in two dimensions. *J. Stat. Phys.* **46**, 507–515.

LANGBEIN W. B., & OTHERS 1947. Topographic characteristics of drainage basins. *U. S. Geol. Surv. Water-Supply Pap.* **968-C**, 125–157.

LENORMAND R. 1985. Différentes mécanismes de déplacements visqueux et capillaries en milieux poreux: Diagramme de phase. *C. R. Acad. Sci. Paris Ser. II* **301**, 247–250.

LENORMAND R., & ZARCONE C. 1985a. Invasion percolation in an etched network: Measurement of a fractal dimension. *Phys. Rev. Lett.* **54**, 2226–2229.

LENORMAND R., & ZARCONE C. 1985b. Two-phase flow experiments in a two-dimensional permeable medium. *Physico-Chemical Hydrodynamics* **6**, 497–506.

LOVEJOY S. 1982. Area-perimeter relation for rain and cloud areas. *Science* **216**, 185–187.

LOVEJOY S., & MANDELBROT B. B. 1985. Fractal properties of rain, and a fractal model. *Tellus* **37A**, 209–232.

LOVEJOY S., & SCHERTZER D. 1985. Generalized scale invariance in the atmosphere and fractal models of rain. *Water Resour. Res.* **21**, 1233–1250.

LOVEJOY S., SCHERTZER D., & TSONIS A. A. 1987. Functional box-counting and multiple elliptical dimensions in rain. *Science* **235**, 1036–1038.

MAHER J. V. 1985. Development of viscous fingering patterns. *Phys. Rev. Lett.* **54**, 1498–1501.

MAKAROV N. G. 1985. On the distortion of boundary sets under conformal mappings. *Proc. London Math. Soc.* **51**, 369–384.

MÅLØY K. J., FEDER J., & JØSSANG T. 1985a. Viscous fingering fractals in porous media. *Phys. Rev. Lett.* **55**, 2688–2691.

MÅLØY K. J., FEDER J., & JØSSANG T. 1985b. Radial viscous fingering in a Hele-Shaw cell. *Report Series, Cooperative Phenomena Project, Department of Physics, University of Oslo,* **85-9**, 1–15.

MÅLØY K. J., BOGER F., FEDER J., JØSSANG T., & MEAKIN P. 1987a. Dynamics of viscous-fingering fractals in porous media. *Phys. Rev. A* **36**, 318–324.

MÅLØY K. J., BOGER F., FEDER J., & JØSSANG T. 1987b. Dynamics and structure of viscous fingers in porous media. in *Time-Dependent Effects in Disordered Materials* (editors R. Pynn & T. Riste, Plenum Press, New York, pp. 111–138).

MÅLØY K. J., FEDER J., & JØSSANG T. 1987c. Viscous fingering in a 2-dimensional strip geometry. *Report Series, Cooperative Phenomena Project, Department of Physics, University of Oslo* **87-10**, 1–11.

MANDELBROT B. B. 1967. How long is the coast of Britain? Statistical self-similarity and fractal dimension. *Science* **155**, 636–638.

MANDELBROT B. B. 1971. A fast fractional Gaussian noise generator. *Water Resour. Res.* **7**, 543–553.

MANDELBROT B. B. 1972. Possible refinement of the lognormal hypothesis concerning the distribution of energy dissipation in intermittent turbulence. in *Statistical Models and Turbulence* (editors M. Rosenblatt & C. Van Atta, Lecture Notes in Physics **12**, Springer, New York, pp. 333–351).

MANDELBROT B. B. 1974. Intermittent turbulence in self-similar cascades: Divergence of high moments and dimension of the carrier. *J. Fluid Mech.* **62**, 331–358.

MANDELBROT B. B. 1975a. *Les Objets Fractals: Forme, Hasard et Dimension.* (Flammarion, Paris).

MANDELBROT B. B. 1975b. Stochastic models of the Earth's relief, the shape and the fractal dimension of the coastlines, and the number-area rule for islands. *Proc. Natl. Acad. Sci. USA* **72**, 3825–3828.

MANDELBROT B. B. 1977. *Fractals: Form, Chance, and Dimension.* (W. H. Freeman, San Fransisco).

MANDELBROT B. B. 1982. *The Fractal Geometry of Nature.* (W. H. Freeman, New York). Page numbers correspond to the 1983 edition.

MANDELBROT B. B. 1983. Fractals in physics: Squig clusters, diffusions, fractal measures, and the unicity of fractal dimensionality. *J. Stat. Phys.* **34**, 895–930.

MANDELBROT B. B. 1985. Self-affine fractals and fractal dimension. *Phys. Scr.* **32**, 257–260.

MANDELBROT B. B. 1986. Self-affine fractal sets. in *Fractals in Physics* (editors L. Pietronero & E. Tosatti, North-Holland, Amsterdam, pp. 3–28).

MANDELBROT B. B. 1987. Fractals. *Encyclopedia of Physical Science and Technology* **5**, 579–593.

MANDELBROT B. B. 1988. *Fractals and Multifractals: Noise, Turbulence and Galaxies.* (Springer, New York)

MANDELBROT B. B., & GIVEN J. A. 1984. Physical properties of a new fractal model of percolation clusters. *Phys. Rev. Lett.* **52**, 1853–1856.

MANDELBROT B. B., & VAN NESS J. W. 1968. Fractional Brownian motions, fractional noises and applications. *SIAM Rev.* **10**, 422–437.

MANDELBROT B. B., & WALLIS J. R. 1968. Noah, Joseph, and operational hydrology. *Water Resour. Res.* **4**, 909–918.

MANDELBROT B. B., & WALLIS J. R. 1969a. Some long-run properties of geophysical records. *Water Resour. Res.* **5**, 321–340.

MANDELBROT B. B., & WALLIS J. R. 1969b. Computer experiments with fractional Gaussian noises. Part 1, Averages and variances. *Water Resour. Res.* **5**, 228–241.

MANDELBROT B. B., & WALLIS J. R. 1969c. Computer experiments with fractional Gaussian noises. Part 2, Rescaled ranges and spectra. *Water Resour. Res.* **5**, 242–259.

MANDELBROT B. B., & WALLIS J. R. 1969d. Computer experiments with fractional Gaussian noises. Part 3, Mathematical appendix. *Water Resour. Res.* **5**, 260–267.

MANDELBROT B. B., & WALLIS J. R. 1969e. Robustness of the rescaled range R/S in the measurement of noncyclic long run statistical dependence. *Water Resour. Res.* **5**, 967–988.

MANDELBROT B. B., PASSOJA D. E., & PAULLAY A. J. 1984. Fractal character of fracture surfaces of metals. *Nature* **308**, 721–722.

MARGOLINA A., NAKANISHI H., STAUFFER D., & STANLEY H. 1984. Monte Carlo and series study of corrections to scaling in two-dimensional percolation. *J. Phys.* A **17**, 1683–1701.

MATSUSHITA M., SANO M., HAYAKAWA Y., HONJO H., & SAWADA Y. 1984. Fractal structures of zinc metal leaves grown by electrodeposition. *Phys. Rev. Lett.* **53**, 286–289.

MATSUSHITA M., HAYAKAWA Y., & SAWADA Y. 1985. Fractal structure and cluster statistics of zinc-metal trees deposited on a line electrode. *Phys. Rev.* A **32**, 3814–3816.

MAULDIN R. D. 1986. On the Hausdorff dimension of graphs and random recursive objects. in *Dimensions and Entropies in Chaotic Systems* (editor G. Mayer-Kress, Springer Verlag, Berlin, pp. 28–33).

MEAKIN P. 1983. Diffusion-controlled cluster formation in 2–6 dimensional space. *Phys. Rev. A* **27**, 1495–1507.

MEAKIN P. 1987a. The growth of fractal aggregates. in *Time-Dependent Effects in Disordered Materials* (editors R. Pynn & T. Riste, Plenum Press, New York, pp. 45–70).

MEAKIN P. 1987b. Scaling properties for the growth probability measure and harmonic measure of fractal strucures. *Phys. Rev. A* **35**, 2234–2245.

MEAKIN P. 1987c. Fractal aggregates and their fractal measures. in *Phase Transitions and Critical Phenomena* (editors C. Domb & J. L. Lebowitz, Academic Press, New York).

MEAKIN P., & WITTEN T. A. 1983. Growing interface in diffusion-limited aggregation. *Phys. Rev. A* **28**, 2985–2989.

MEAKIN P., STANLEY H. E., CONIGLIO A., & WITTEN T. A. 1985. Surfaces, interfaces, and screening of fractal structures. *Phys. Rev. A.* **32**, 2364–2369.

MEAKIN P., CONIGLIO A., STANLEY H. E., & WITTEN T. A. 1986. Scaling properties for the surfaces of fractal and nonfractal objects: An infinite hierarchy of critical exponents. *Phys. Rev. A* **34**, 3325–3340.

MENEVEAU C., & SREENIVASAN K. R. 1987. Simple multifractal cascade model for fully developed turbulence. *Phys. Rev. Lett.* **59**, 1424–1427.

MILLER G. S. P. 1986. The definition and rendering of terrain maps. *Computer Graphics* **20**, 39–48.

MURAT M., & AHARONY A. 1986. Viscous fingering and diffusion-limited aggregates near percolation. *Phys. Rev. Lett.* **57**, 1875–1878.

NIEMEYER L., PIETRONERO L., & WIESMANN H. J. 1984. Fractal dimension of dielectric breakdown. *Phys. Rev. Lett.* **52**, 1033–1036.

NITTMANN J., & STANLEY H. E. 1986. Tip splitting without interfacial tension and dendritic growth patterns arising from molecular anisotropy. *Nature* **321**, 663–668.

NITTMANN J., DACCORD G., & STANLEY H. E. 1985. Fractal growth of viscous fingers: Quantitative characterization of a fluid instability phenomenon. *Nature* **314**, 141–144.

NITTMANN J., STANLEY H. E., TOUBOUL E., & DACCORD G. 1987. Experimental evidence for multifractality. *Phys. Rev. Lett.* **58**, 619.

OPPENHEIMER P. E. 1986. Real time design and animation of fractal plants and trees. *Computer Graphics* **20**, 55–64.

OXAAL U., MURAT M., BOGER F., AHARONY A., FEDER J., & JØSSANG T. 1987. Viscous fingering on percolation clusters. *Nature* **329**, 32–37.

PATERSON L. 1981. Radial fingering in a Hele-Shaw cell. *J. Fluid Mech.* **113**, 513–529.

PATERSON L. 1984. Diffusion-limited aggregation and two-fluid displacements in porous media. *Phys. Rev. Lett.* **52**, 1621–1624.

PATERSON L., HORNHOF V., & NEALE G. 1982. A consolidated porous medium for the visualization of unstable displacements. *Powder Technol.* **33**, 265–268.

PEITGEN H.-O. 1988. *The Art of Fractals, A Computer Graphical Introduction.* (To be published, Springer-Verlag, Berlin).

PEITGEN H.-O. & RICHTER P. H. 1986. *The Beauty of Fractals.* (Springer-Verlag, Berlin).

PFEIFER P., & AVNIR D. 1983. Chemistry in noninteger dimensions between two and three. I. Fractal theory and heterogeneous surfaces. *J. Chem. Phys.* **79**, 3558–3565. Erratum: *J. Chem. Phys.* (1984) **80**, 4573.

PFEIFER P., AVNIR D., & FARIN D. 1983. Ideally irregular surfaces, of dimension greater than two, in theory and practice. *Surface Sci.* **126**, 569–572.

PFEIFER P., AVNIR D., & FARIN D. 1984. Scaling behavior of surface irregularity in the molecular domain: From adsorbtion studies to fractal catalysts. *J. Stat. Phys.* **36**, 699–716.

PIETRONERO L. & TOSATTI E., *editors,* 1986. *Fractals in Physics.* (North-Holland, Amsterdam).

PIKE R., & STANLEY H. E. 1981. Order propagation near the percolation threshold. *J. Phys. A* **14**, L169–L177.

PROCACCIA I. 1986. The characterization of fractal measures as interwoven sets of singularities: Global universality at the transition to chaos. in *Dimensions and Entropies in Chaotic Systems* (editor G. Mayer-Kress, Springer-Verlag, Berlin, pp. 8–18).

PYNN R. & RISTE T., *editors.* 1987. *Time-Dependent Effects in Disordered Materials.* (Plenum Press, New York).

PYNN R. & SKJELTORP A., *editors.* 1985. *Scaling Phenomena in Disordered Systems.* (Plenum Press, New York).

RAMMAL R., TANNOUS C., BRETON P., & TREMBLAY A. M. S. 1985. Flicker $(1/f)$ noise in percolation networks: A new hierarchy of exponents. *Phys. Rev. Lett.* **54**, 1718–1721.

REYNOLDS P. J., KLEIN W., & STANLEY H. E. 1977. A real-space renormalization group for site and bond percolation. *J. Phys. C* **10**, L167–L172.

ROBERTS J. N. 1986. Comment about fractal sandstone pores. *Phys. Rev. Lett.* **56**, 2111.

ROJANSKI D., HUPPERT D., BALE H. D., DACAI X., SMITH P. W., FARIN D., SERI-LEVY A., & AVNIR D. 1986. Integrated fractal analysis of silica: Adsorbtion, electronic energy transfer, and small-angle X-ray scattering. *Phys. Rev. Lett.* **56**, 2505–2508.

ROSSO M., GOUYET J. F., & SAPOVAL B. 1985. Determination of percolation probability from the use of a concentration gradient. *Phys. Rev. B* **32**, 6053–6054.

ROSSO M., GOUYET J. F., & SAPOVAL B. 1986. Gradient percolation in three dimensions and relation to diffusion fronts. *Phys. Rev. Lett.* **57**, 3195–3198.

RYS F. S., & WALDVOGEL A. 1986. Fractal shape of hail clouds. *Phys. Rev. Lett* **56**, 784–787.

SAFFMAN P. G., & TAYLOR G. I. 1958. The penetration of a fluid into a medium or Hele-Shaw cell containing a more viscous liquid. *Proc. R. Soc. Lond.* **245**, 312–329.

SALEUR H., & DUPLANTIER B. 1987. Exact determination of the percolation hull exponent in two dimensions. *Phys. Rev. Lett.* **58**, 2325–2328.

SAPOVAL B., ROSSO M., & GOUYET J. F. 1985. The fractal nature of a diffusing front and the relation to percolation. *J. Phys. Lett.* **46**, L149–L156.

SAPOVAL B., ROSSO M., GOUYET J. F., & COLONNA J. F. 1986. Dynamics of the creation of fractal objects by diffusion and $1/f$ noise. *Solid State Ionics* **18**, 21–30.

SAYLES R. S., & THOMAS T. R. 1978a. Surface topography as a nonstationary random process. *Nature* **271**, 431–434.

SAYLES R. S., & THOMAS T. R. 1978b. Reply to 'Topography of random surfaces' by M. V. Berry and J. H. Hannay (1978). *Nature* **273**, 573.

SCHAEFER D. W., MARTIN J. E., WILTZIUS P., & CANNELL D. S. 1984. Fractal geometry of colloidal aggregates. *Phys. Rev. Lett.* **52**, 2371–2374.

SHANTE V. K. S., & KIRKPATRICK S. 1971. An introduction to percolation theory. *Adv. Phys.* **20**, 325–357.

SHAW T. M. 1987. Drying as an immiscible displacement process with fluid counterflow. *Phys. Rev. Lett.* **59**, 1671–1674.

SINHA S. K., FRELTOFT T., & KJEMS J. 1984. Observation of power-law correlations in silica-particle aggregates by small angle neutron scattering. in *Aggregation and Gelation* (editors F. Family & D. P. Landau, North-Holland, Amsterdam, pp. 87–90).

SKAL A. S., & SHLOVSKII B. I. 1975. Topology of an infinite cluster in the theory of percolation and its relationship to the theory of hopping conduction. *Sov. Phys. Semicond.* **8**, 1029–1032.

SØRENSEN P. R. 1984. Fractals: Exploring the rough edges between dimensions. *Byte* **9**, 157–172.

STANLEY H. E. 1977. Cluster shapes at the percolation threshold: An effective cluster dimensionality and its connection with critical point exponents. *J. Phys. A* **10**, L211–L220.

STANLEY H. E. 1985. Fractal concepts for disordered systems: The interplay of physics and geometry. in *Scaling Phenomena in Disordered Systems* (editors R. Pynn & A. Skjeltorp, Plenum Press, New York, pp. 49–69).

STANLEY H. E. & OSTROWSKY N., *editors*. 1985. *On Growth and Form.* (Martinus Nijhoff, Dordrecht).

STAUFFER D. 1979. Scaling theory of percolation clusters. *Phys. Rep.* **54**, 1–74.

STAUFFER D. 1985. *Introduction to Percolation Theory.* (Taylor & Francis, London).

STAUFFER D. 1986. Percolation and cluster size distribution. in *On Growth and Form* (editors H. E. Stanley & N. Ostrowsky, Martinus Nijhoff, Dordrecht, pp. 79–100).

STOKES J. P., WEITZ D. A., GOLLUB J. P., DOUGHERTY A.,ROBBINS M. O., CHAIKIN P. M., & LINDSAY H. M. 1986. Interfacial stability of immiscible displacement in a porous medium. *Phys. Rev. Lett.* **57**, 1718–1721.

SYKES M. F., & ESSAM J. W. 1964. Exact critical percolation probabilities for site and bond percolation in two dimensions. *J. Math. Phys.* **5**, 1117–1127.

VAN MEURS P. 1957. The use of transparent three-dimensional models for studying the mechanism of flow processes in oil reservoirs. *Trans. AIME* **210**, 295–301.

VAN MEURS P., & VAN DER POEL C. 1958. A theoretical description of water-drive processes involving viscous fingering. *Trans. AIME* **213**, 103–112.

VON SMOLUCHOWSKI M. 1916. Drei Vorträge über Diffusion, Brownsche Molekularbewegung und Koagulation von Kolloidteilchen. *Phys. Z.* **17**, 557–599.

VOSS R. F. 1984. The fractal dimension of percolation cluster hulls. *J. Phys. A* **17**, L373–L377.

VOSS R. F. 1985a. Random fractals: Characterization and measurement. in *Scaling Phenomena in Disordered Systems* (editors R. Pynn & A. Skjeltorp, Plenum Press, New York, pp. 1–11).

VOSS R. F. 1985b. Random fractal forgeries. in *Fundamental Algorithms in Computer Graphics* (editor R. A. Earnshaw, Springer-Verlag, Berlin, pp. 805–835. Color plates on pp. 13–16).

WEITZ D. A., & HUANG J. S. 1984. Self-similar structures and the kinetics of aggregation of gold colloids. in *Aggregation Gelation* (editors F. Family & D. P. Landau, North-Holland, Amsterdam, pp. 19–28).

WEITZ D. A., & OLIVERIA M. 1984. Fractal structures formed by kinetic aggregation of aqueous gold colloids. *Phys. Rev. Lett.* **52**, 1433–1436.

WEITZ D. A., LIN M. Y., HUANG J. S., WITTEN T. A., SINHA S. K., GERTNER J. S., & BALL C. 1985. Scaling in colloid aggregation. in *Scaling Phenomena in Disordered Systems* (editors R. Pynn & A. Skjeltorp, Plenum Press, New York, pp. 171–188).

WIENER N. 1923. Differential-space. *J. Math. Phys. Mass. Inst. Technol.* **2**, 131–174.

WILKINSON D., & WILLEMSEN J. F. 1983. Invasion percolation: A new form of percolation theory. *J. Phys. A* **16**, 3365–3376.

WITTEN T. A., & SANDER L. M. 1983. Diffusion-limited aggregation. *Phys. Rev. B* **27**, 5686–5697.

WONG P.-Z. 1985. Scattering by homogeneous systems with rough internal surfaces: Porous solids and random-field Ising systems. *Phys. Rev. B* **32**, 7417–7424.

WONG P.-Z., HOWARD J., & LIN J.-S. 1985. Surface roughening and the fractal nature of rocks. Schlumberger-Doll Research Preprint.

YOUNG A. P., & STINCHCOMBE R. B. 1975. A renormalization group theory for percolation problems. *J. Phys. C* **8**, L535–L540.

ZIFF R. M. 1986. Test of scaling exponents for percolation-cluster perimeters. *Phys. Rev. Lett.* **56**, 545–548.

Author Index

Subject Index